彩图 2-1　722 型分光光度计

U0161547

彩图 2-2　T9 紫外可见分光光度计

彩图 2-3　LS 45 荧光分光光度计

彩图 2-4　Nicolet iS50 傅里叶红外光谱仪

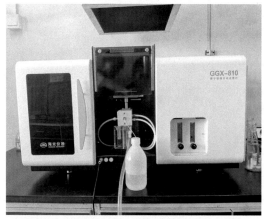

彩图 2-5　GGX-810 原子吸收分光光度计

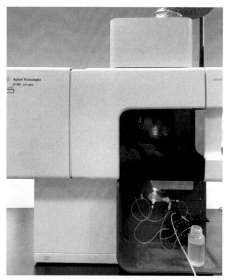

彩图 2-6　5100 电感耦合等离子原子发射光谱仪

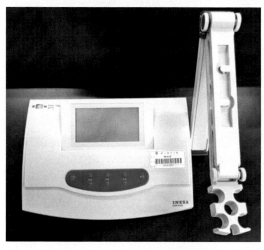

彩图 2-7　pHS-3C 型酸度计

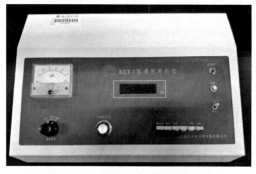

彩图 2-8　KLT-1 型通用库仑仪

彩图 2-9　CHI660E 电化学工作站

彩图 2-10　Titrando 888 型自动电位滴定仪

彩图 2-11　7820A 气相色谱仪

彩图 2-12　LC-20AD 高效液相色谱仪

彩图 2-13
BECKMAN P/ACE MDQ 高效毛细管电泳仪

彩图 2-14　Metrohm 861 型离子色谱仪

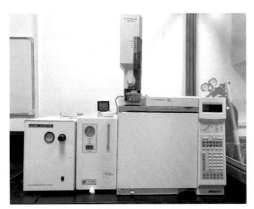

彩图 2-15
7890B-5977B GC-MSD 气质联用仪

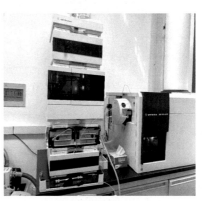

图 2-16
6230 飞行时间液质联用系统（TOF LC/MS）

高等教育新形态教材

Instrumental Analysis Experiments

仪器分析实验

（英汉双语版）

翁雪香 主编

吴 靓 张 露 副主编

化学工业出版社

·北京·

内容简介

中英文双语的《仪器分析实验》新形态教材以浙江省线下一流本科课程"仪器分析"为依托，融合了纸质教材和多介质数字化资源。其中纸质教材内容包括光谱分析、电化学分析、分离分析。数字资源包括经典仪器分析实验视频、中英双语的仪器分析教学课件、课后习题及虚拟仿真实验等线上学习内容。

本书希望为仪器分析学习者提供一站式学习服务，力求将理论知识和实验操作结合起来，真正实现理论与实验课程的有机统一。本书可作为高等院校的仪器分析实验教材。

图书在版编目（CIP）数据

仪器分析实验：英汉对照/翁雪香主编；吴靓，张露副主编. —北京：化学工业出版社，2023.9（2024.11重印）
ISBN 978-7-122-43680-1

Ⅰ.①仪… Ⅱ.①翁… ②吴… ③张… Ⅲ.①仪器分析-实验-高等学校-教材-英、汉 Ⅳ.①O657-33

中国国家版本馆 CIP 数据核字（2023）第 111409 号

责任编辑：李　丽　　　　　　　文字编辑：曹　敏
责任校对：王鹏飞　　　　　　　装帧设计：关　飞

出版发行：化学工业出版社
　　　　　（北京市东城区青年湖南街 13 号　邮政编码 100011）
印　　装：北京虎彩文化传播有限公司
787mm×1092mm　1/16　印张 13½　彩插 1　字数 329 千字
2024 年 11 月北京第 1 版第 2 次印刷

购书咨询：010-64518888　　　　售后服务：010-64518899
网　　址：http://www.cip.com.cn

定　　价：　69.00 元

《仪器分析实验》（英汉双语版）
编写人员名单

主　　编　翁雪香

副主编　吴　靓　张　露

编写人员　（按姓名汉语拼音排序）

冯九菊　　康晋伟　　孔黎春

刘卫东　　罗　芳　　钱兆生

王卫平　　翁雪香　　吴　靓

袁军华　　袁培新　　张　露

前言

随着信息技术与教育教学的不断融合深入，信息化教学作为一种全新的教学方式，对教学的重要载体——教材的内容和功能提出了新的要求。新形态教材是"纸质+数字化资源"教材，是在传统教材基础上融合了人工智能、虚拟现实以及移动互联网方面一项或多项媒介的多介质教材。

仪器分析实验是各大专院校化学、医药、环境等相关专业的一门重要的专业核心课程，一般与仪器分析理论课程同步开设。通过该课程的学习，学生能加深对有关仪器分析方法基本原理的理解；学会正确地使用分析仪器；能根据实际样品的分析要求，合理地选择实验条件；能正确处理数据和表达实验结果。更为重要的是通过实验课程的实践学习，能培养学生严谨求是的科学态度、勇于创新和敢于质疑的思维品质。

全球经济的一体化，培养学生的国际视野和国际交流合作能力尤为重要，目前双语教育在各高校逐步开展。分析仪器飞快迭代，新的仪器和分析方法层出不穷。为更好地促进信息革命时代学生在仪器分析实验课程中达成学习目标，本教材以高等学校化学化工专业本科生仪器分析实验教学大纲为依据，结合所在学校的仪器条件和几十年的教学实践经验，借鉴了国内很多前辈编写的教材及一些国外仪器分析实验手册，编写了43个中、英文对照，内容涵盖光、电、色谱领域的经典及拓展实验，实验内容由浅至深，由基础至综合。本书编写中注重理论和实践相结合、注重经典和现代的结合。

本教材以二维码的方式嵌入了其中的8个经典仪器分析实验视频，还有配套的仪器分析理论和实验课程线上学习资源，包括每个章节的学习目标、中英双语版的教学课件、课后习题、虚拟仿真实验等。丰富的线上学习资源旨在培养学生通过预习和复习而获得新知的能力。

纸质教材共分三章，第一章，第二章中的电化学仪器的结构及使用和第三章中的电化学实验由张露负责翻译；第二章中光谱分析仪器的结构及使用、第三章中的部分光学实验（实验9~12，14和15）由吴靓负责编译；实验1、3、8、13、22、26、27、29、31、33分别由吴靓、刘卫东、钱兆生、张露、袁培新、冯九菊、袁军华和王卫平负责原始素材的搜集和翻译；翁雪香负责全书素材的收集、其余实验的编译及全文校对统稿。此外，康晋伟为第二章仪器的结构及使用提供了大量有益的参考资料，并负责这些仪器照片的拍摄工作；孔黎春为傅里叶红外光谱仪的结构及使用提供了原始素材，并对红外实验的选择提出了宝贵的建议；罗芳提供了气质联用仪、液质联用仪结构及使用的原始素材。前期资料搜集和后期编排工作主要由分析化学专业研究生余丽圆、张娟华，化学教育研究生李莎、郑笑雅、张雨琦、王岳、赵御彤、吴焕焕同学负责完成。在此向他们表示衷心的感谢！

本书的出版得到了浙江师范大学 2020 年实验教学示范中心软件建设项目和浙江师范大学化学与材料科学学院的大力支持，在此也表示衷心的感谢！

由于编者水平有限，书中难免有一些疏漏，恳请读者批评指正！

编者

2023 年 5 月

目录

Chapter 1　Fundamentals of Instrumental Analysis Experiments —— 89

Chapter 2　The Structure and Operation of Instruments —— 94

Chapter 3　Instrumental Analysis Experiments —— 114

第一章 仪器分析实验的基本知识

一、实验室安全知识

（1）在任何实验室中，安全都是最重要的。进入实验室时，请确认安全淋浴、洗眼器和灭火器的位置。请佩戴防护眼镜，穿上白大褂，无一例外。

（2）所有化学废物都应装在贴有适当标签的废物容器中，确保丢废弃物的时候标签是正确的。废物如果混合不当，可能会发生事故。

二、仪器分析实验的基本要求

仪器分析实验是仪器分析课程的重要内容，一般与理论课同步开设。它是学生在教师指导下，以分析仪器为工具，亲自动手获得所需物质化学组成、结构及含量等信息的教学实践活动。通过仪器分析实验课程的学习，学生能加深对有关仪器分析方法基本原理的理解，掌握仪器分析实验的基本知识和技能；学会正确地使用分析仪器；能根据实际样品的分析要求，合理地选择实验条件；能正确处理数据和表达实验结果。通过实验课程的实践学习，培养学生严谨求是的科学态度、勇于创新和敢于质疑的思维品质。为了达到以上学习目标，对仪器分析实验提出以下基本要求：

（1）**充分做好预习工作** 仪器分析所用的仪器一般较昂贵，同一实验室不可能购置多套同类仪器，仪器分析实验通常都采用大循环方式组织教学。许多情况下，理论课的内容还未学习，实验已经开始。因此，学生在实验前必须做好预习工作，仔细阅读仪器分析实验教材和理论课程教材，明确实验目的，透彻理解实验的相关原理；认真学习线上相关实验视频，了解仪器的操作规程，并写好预习报告。

（2）**正确使用仪器** 要在教师指导下使用仪器，未经教师允许不得随意开动或关闭仪器，更不得随意旋转仪器旋钮、改变仪器工作参数等。实验中如发现仪器工作不正常，应及时报告教师处理。

（3）**培养良好的实验习惯** 实验过程中，细心观察实验现象，仔细记录实验条件，用心分析测试的原始数据。实验结束后，应将所用仪器复原。及时洗涤用过的器皿，始终保持实验室的整洁有序。

（4）**写好实验报告** 实验报告应简明扼要，图表清晰。实验报告的内容包括实验名称、完成日期、实验目的、方法原理、仪器及试剂、主要实验步骤、实验数据或图谱、实验数据分析和结果处理、问题讨论等。

三、实验数据处理和结果表达

（一）评价分析方法和分析结果的基本指标

定量分析是仪器分析的主要任务之一。评价定量分析方法及其分析结果需要一定的性能参数和指标。一般来说，定量分析方法具有以下一些常用性能参数与指标。

（1）标准曲线及其线性范围　定量分析普遍使用的方法是标准曲线法。标准曲线（standard curve）又称校准曲线（calibration curve），是指被测物质的浓度（或含量）与仪器响应信号的关系曲线。

标准曲线的直线部分所对应的被测物质浓度（或含量）的范围称为该方法的线性范围。一般来说，分析方法的线性范围越宽越好。

标准曲线是依据标准系列的浓度（或含量）和相应的响应信号测量值来绘制的。由于存在着随机误差，即单次测量值（x 或 y）与 n 次测量平均值（$\overline{x}$ 或 $\overline{y}$）存在着平均偏差 $\overline{d}$。根据最小二乘法原理，研究因变量与自变量之间关系的方法称为回归分析。如果只有一个自变量，称为一元线性回归分析法。设浓度分别为 x_1，x_2，…，x_i，…，x_n 的标准系列，其响应信号的测量值为 y_1，y_2，…，y_i，…，y_n。如果各点对某一直线的偏差平方和 $\sum d^2$ 为最小或为零，则该直线即为最佳一元回归直线。根据这一原理，设一元线性方程

$$y = bx + a \tag{1-1}$$

求解式(1-1)一元回归线性方程，得

$$b = \frac{\sum_{i=1}^{n} (x_i - \overline{x})(y_i - \overline{y})}{\sum_{i=1}^{n} (x_i - \overline{x})^2} \tag{1-2}$$

$$a = \overline{y} - b\overline{x} \tag{1-3}$$

其中，$\overline{x} = \sum_{i=1}^{n} x_i / n$，$\overline{y} = \sum_{i=1}^{n} y_i / n$。由式(1-1)可知，当 $x = 0$ 时，$y = a$；当 $x = \overline{x}$，$y = \overline{y}$。过（0，a）和（$\overline{x}$，$\overline{y}$）两点在 x 浓度范围内作直线，此直线就是给定的数据组（x_i，y_i）所确定的一条最佳标准曲线。

判断此标准曲线线性关系是否成立，具有实际意义，可用相关系数 r 来检验，r 是表征变量之间相关程度的一个统计参数见式(1-4)。

$$r = \pm \frac{\sum_{i=1}^{n} (x_i - \overline{x})(y_i - \overline{y})}{\left[\sum_{i=1}^{n} (x_i - \overline{x})^2 \sum_{i=1}^{n} (y_i - \overline{y})^2 \right]^{1/2}} \tag{1-4}$$

r 值在 $-1 \sim +1$ 之间，当 $|r| = 1$ 时，y 与 x 之间存在着严格的线性关系，所有 y 值都在一条直线上；当 $r = 0$ 时，y 与 x 之间不存在线性关系；当 $0 < |r| < 1$ 时，y 与 x 之间有一定的线性关系。$|r|$ 越接近 1，y 与 x 的相关性越好。

（2）检出限和灵敏度　分析方法的检出限是指能以适当的置信水平（通常取置信水平 99.7%）检测出被测组分的最低浓度或最小质量。检出限（LOD）由最低检测信号值与空白噪声计算［式(1-5)］，最低检出浓度和最小检出质量的单位分别用 $\mu g/mL$，ng/mL 和 μg，ng，pg 表示。

$$LOD = \frac{X_L - \overline{X}_b}{S} = \frac{3s_b}{S} \tag{1-5}$$

式中，X_L 是可被检测的最小分析信号值；$\overline{X_b}$ 是对空白进行多次测量所得空白信号平均值；s_b 为空白信号的标准偏差；S 是低浓度区标准曲线的斜率，它表示被测组分浓度改变一个单位时分析信号的变化程度，即分析方法的灵敏度。

在仪器分析中，分析方法的灵敏度直接依赖于检测器的灵敏度与仪器的放大倍数。随着灵敏度的提高，噪声也随之增大，而信噪比 S/N 和分析方法的检测能力不一定会改善和提高。如果只给出灵敏度，而不给出获得此灵敏度的仪器条件，则各分析方法之间的检测能力没有可比性。由于灵敏度没有考虑到测量噪声的影响，因此，现在已不用灵敏度而推荐用检出限来表征分析方法的检测能力。

(3) 准确度 准确度是指在一定实验条件下测定值 x 与真值或标准值 μ 符合的程度 [式(1-6)]。它表征系统误差的大小，以误差或相对误差 E_r 表示。误差或相对误差越小，准确度越高。

$$E_r = \frac{x - \mu}{\mu} \times 100\% \tag{1-6}$$

在实际工作中，通常用标准物质或标准方法进行对照试验。在无标准物质或标准方法时，常用加入被测定组分的纯物质进行回收试验来估计与确定准确度，即测定其回收率。值得注意的是，用回收率来估计测定的准确度，只适用于系统误差随浓度改变的情况。

在误差较小时，多次平行测定的平均值 $\overline{x}$ 接近于真值 μ，故在实际工作中常将 $\overline{x}$ 作为 μ 的估计值使用。

(4) 精密度 精密度是指使用同一方法，对同一试样进行多次平行测定所得测定值彼此间相符合的程度，表征测定过程中随机误差的大小，又称重复性，常用标准偏差 s 或相对标准偏差 s_r 表示，其数学表达式为

$$s = \left[\frac{1}{n-1} \sum (x_i - \overline{x})^2 \right]^{1/2} \tag{1-7}$$

$$s_r = \frac{s}{\overline{x}} \times 100\% \tag{1-8}$$

式中，x_i 是单次测定值；$\overline{x}$ 是 n 次测定的平均值；n 是重复测定次数。

(5) 选择性 选择性（selectivity）是指用某种分析方法测定某组分时能够避免试样中其他共存组分干扰的能力。选择性通常表示为在指定的测量准确度下，共存组分的允许量（浓度或质量）与待测组分的量（浓度或质量）的比值。该比值越大，表明在指定的准确度下，该方法的抗干扰能力越强，即选择性越好。

(6) 响应时间和分析效率 某分析仪器的响应时间（response time）是指激发信号刺激试样而使仪器检测信号达到总变化量一定百分数所需要的时间。例如，某离子选择性电极的响应时间是指离子选择性电极与参比电极从接触试液开始到电极电位变化稳定（±1mV）所需要的时间。一般来说，响应时间越短越好。

分析效率（速度）是在单位时间内能够测定试样的个数。一般来说，分析效率越高越好。

（二）分析数据和结果的表达

(1) 测量值的读数与表达 在仪器分析中，一般都是仪器把与化学信息有关的原始信号转换成电信号，经放大由仪器用数字直接显示。为保证测量的准确性，对显示出的信号必须

正确读数。

（2）**分析数据和结果的表达**　分析数据和分析结果的表示法主要有列表法和数学方程表示法，其基本要求是准确、清晰和便于应用。

① 列表法。列表法是以表格形式表示数据，直观、简明，记录实验数据多用此法。列表需标明表名，表的纵列一般为试验编号或因变量，横列为自变量。行首或列首应写上名称及量纲。名称尽量用符号表示，单位和符号之间用斜线分隔开，如该列数据表示温度 T，则该列首应写成"T/K"。记录数据应符合有效数字的规定。书写时应整齐统一，小数点要上下对齐，以利于数据的比较分析。表中的某个数据需要特殊说明时，可在数据上角作一标记，如①，在表的下方加注说明。

② 数学方程表示法。在仪器分析中，绝大多数情况下都是相对测量，需用标准曲线进行定量分析，由于测量误差不可避免，所有的数据点都处在同一条直线上是不多见的。特别是测量误差较大时，用简单的方法很难绘出合理的标准曲线。这种情况下以数学方程表示法来描述自变量与因变量之间的关系较为妥当。如最佳一元回归直线及一元线性方程可用作图软件如 EXCEL 或 ORIGIN 完成。

第二章 仪器的结构及使用

一、光谱分析仪器的结构及使用

(一) 722 型分光光度计

(1) 性能与结构 722 型分光光度计是在可见光谱区域内使用的一种单光束型仪器，工作波长范围为 360~800nm。它采用钨丝白炽灯光源，棱镜单色器，自准式光路，用 GD-7 型真空光电管为光电转换器，以场效应管为放大器，微电流用微安表显示。722 型分光光度计如图 2-1 所示。

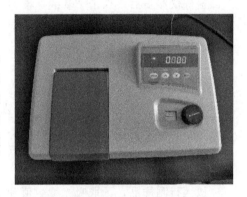

图 2-1　722 型分光光度计（彩图）

(2) 仪器操作步骤

① 开启电源，预热仪器，选择检测 MODE 为 "T" 模式。

② 打开样品室盖（光门自动关闭），按 "0％T" 键，使数字显示 "0.000"。

③ 旋动仪器波长手轮，把测试所需的波长调节至刻度线处。

④ 选择测试用的比色皿，把盛放参比和待测的样品放入样品架内，通过样品架拉杆来选择样品的位置。

⑤ 盖上样品室盖，将装有参比溶液的比色皿拉入光路，按 "100％T" 键，使数字显示为 100.0。

⑥ 将被测溶液置于光路中，数字表上直接读出被测溶液的透光率（T）值。

⑦ 对于吸光度 A 的测量，参照第②、⑤步调整仪器的 "0.000" 和 "100.0"。选择测量模式 A，使数字显示为 "0.000"。将待测液移入光路，显示值即为试样的吸光度 A 值。

(3) 注意事项

① 为使仪器内部达到热平衡，开机预热时间不小于 30min。

② 若连续测定时间过长，光电管会疲劳造成读数漂移。因此，每次读数后应随手打开试样室盖（光闸自动关闭）。

(二) T9 紫外可见分光光度计

(1) 性能与结构 T9 双光束紫外可见分光光度计（UV-Vis）有特制光栅、超低噪声信号检测系统。吸光度范围宽达 -6~6，充分满足高吸光度样品的测试需求。它采用分立式三

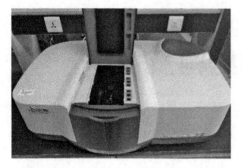

图 2-2 T9 紫外可见分光光度计（彩图）

缝组合连续可变狭缝设计，可自动在 0.1～5.0nm 范围内进行光谱带宽扫描，并识别样品分子共振吸收最强时的光谱带宽，从而确定正确的实验条件。T9 紫外可见分光光度计如图 2-2 所示。

（2）仪器操作步骤

① 开机　开启仪器，打开电脑，双击电脑桌面软件 "UVWIN T9CS"，等待仪器初始化自检。

② 吸收光谱扫描

a. 在仪器工作界面上点击 "光谱扫描"，设置扫描的波长范围后，点击 "确定"。

b. 在两个配对比色皿中都装入空白溶液，透光表面擦拭干净后，放入吸收池中。

c. 盖好仪器盖后，点击 "基线" 按钮，开始自动校准。

d. 基线校正完成后，参比池不变，试样池换为待测液，点击 "开始" 按钮，扫描谱图。

e. 记录最大吸收波长。

f. 导出数据，保存。

③ 标准曲线的绘制和未知样的测定

a. 点击 "定量测定"，设置参数（输入定量测定波长和重复次数）。

b. 点击 "开始"，输入样品的浓度值，仪器自动给出吸光度值。

c. 依次更换不同浓度的标准溶液，点击 "开始"，读取吸光度数值。

d. 系统自动显示标准曲线。

e. 放入待测样品，读取吸光度数值和相应的浓度值。

④ 关机　保存数据，取出比色皿，退出软件，关机。

（3）注意事项

① 在仪器进行自检的时候，不要对仪器进行任何操作。

② 在基线校正的时候，确保试样和参比光束上无任何障碍物，并且试样室中无试样。

③ 更换试样时，比色皿需用待测溶液润洗 2～3 次。

（三）LS 45 荧光分光光度计

（1）性能与结构　LS 45 荧光分光光度计可测定荧光、磷光、生物发光或化学发光。其激发狭缝为 2.5～15nm，发射狭缝为 2.5～20nm。该仪器采用脉冲式氙灯，其寿命长、电源供应简单，也不需长时间预热，可大大减少光解作用。用软件控制即可测定磷光，不需附件。LS 45 荧光分光光度计如图 2-3 所示。

图 2-3　LS 45 荧光分光光度计（彩图）

（2）仪器操作步骤

① 溶液配制。按要求配制系列标准溶液和待测液。

② 开机。开启仪器，打开电脑，打开软件。

③ 激发光谱和发射光谱绘制

a. 放入参比溶液，单击 "测定"，点击 "自动清零"。

b.放入其中一个标准溶液,设定发射波长,在特定波长范围内扫描该标准溶液的激发光谱。

c.设定激发波长,在特定波长范围内扫描标准溶液的发射光谱。

④ 定量测定

a.设置合适的参数。

b.依次测定系列标准溶液和被测样品的荧光强度。

c.保存数据。

d.绘制荧光强度对浓度的标准曲线,并通过线性拟合方程计算被测物的含量。

⑤ 关机。退出测试系统,待仪器散热 30min 后关闭仪器和计算机。

(四) Nicolet iS50 傅里叶红外光谱仪

(1) 性能与结构 Nicolet iS50 傅里叶红外光谱仪 (图 2-4) 是配有专用附件和集成软件的一体式材料分析工作站,其 Vectra 系列干涉仪高精度动镜定位技术突破了传统气动或机械式干涉仪的性能极限,不受振动和噪声的影响,为快速连续扫描、步进扫描和双通道检测等各项应用提供了最佳光谱检测技术。

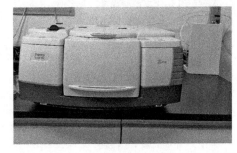

图 2-4 Nicolet iS50 傅里叶红外光谱仪 (彩图)

(2) 仪器操作步骤

① 打开仪器电源。

② 打开计算机,当光学台自诊断完成后双击 OMNIC 图标。

③ 点击 Experiment Setup (实验设置),点击 Collect 采集数据。

a.点击 Bench tab (平台参数) 查看干涉信号的强度,应为 6V 左右。

b.点击 Collect,选择背景采集模式 (采集背景方式通常选第一或第三),然后点击 OK。

④ 采集光谱

a.当采集背景方式为第三种时,将空白 KBr 压片放入样品架,置入样品仓。点击背景采集。

b.将含样品的 KBr 片放入样品架,置入样品仓。点击样品采集。在出现的对话框中输入文件名,点击 "OK,OK",待系统执行采集后,左上角出现的窗口中选 "YES",该光谱即在显示窗口中出现。

c.当背景方式为第一种时,背景与样品的采集依屏幕提示操作即可。

⑤ 谱图处理

a.转到 "A" 方式点击 Aut Bsln 进行自动基线校正后返回 "T" 方式。

b.点击 Find.pks (标峰),之后点击 Replace 替换原图。

c.点击 Print 打印光谱。

d.点击 File (文件)、Save As (另存为),输入文件名,然后 "保存"。

e.点击 Clear 清除不需要的光谱。

(3) 注意事项

① 仪器干涉仪密封仓、样品仓内应置充分、有效的干燥剂 (变色硅胶的颜色不得转红)。

② 所有接入插头的变更，均需在切断电源状态下进行。

③ 确保整个系统须可靠接地，若电源波动较大，请加接净化稳压电源。断电后须人工复位。

④ 保持工作环境温度在 15～25℃ 之间，相对湿度<60％。

（五）GGX-810 原子吸收分光光度计

（1）性能与结构 GGX-810 原子吸收分光光度计（AAS）采用 8 灯立式转塔系统，自动旋转，切换元素灯后可快速开展分析工作（图 2-5）。该仪器能实现波长自动定位，定位精度<±0.1nm。而且它采用浮动光学平台设计，这保证了光学系统高稳定性，减少震动、机械应力变化的影响。

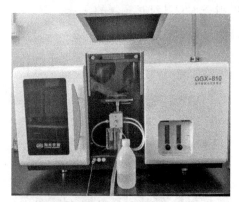

图 2-5　GGX-810 原子吸收分光
光度计（彩图）

（2）仪器操作步骤

① 开机

a. 打开电脑，打开空气压缩机，打开仪器开关。

b. 打开仪器自带软件 "AAS. exe"。

② 标准溶液及未知样品的吸光度测定

a. 新建文件名，选择待测元素、定峰、自动灯位、清零。

b. 打开方法库，打开设置好的 "被测元素"。

c. 打开乙炔，点火。

d. 点击 "分析测试"，插入质控样（质量控制样品）。改测定次数为 "3"。

e. 将去离子水喷入仪器，点击 "标准空白"，点击 "测量"，仪器自动清零。

f. 依次喷入不同浓度的标准溶液（从稀至浓），测量记录其吸光度数值。

g. 喷入去离子水，清洗气路。

h. 喷入未知样品，记录其吸光度值。

i. 测试完毕，喷入去离子水，清洗气路。

③ 关机

a. 关闭测试用空心阴极灯，关闭乙炔钢瓶总阀，待管内乙炔燃尽，关闭乙炔气通阀，关闭空气压缩机。

b. 退出软件，关机。

（3）注意事项

① 该仪器应放在干燥的房间内，使用时放置在坚固平稳的工作台上，室内照明不宜太强。热天时不能用电扇直接向仪器吹风，防止灯泡灯丝发亮不稳定。

② 使用本仪器前，使用者应先了解本仪器的结构和工作原理，以及各个操纵旋钮的功能。

③ 在仪器尚未接通电源时，电表指针必须于 "0" 刻线上，否则可用电表上的校正螺丝进行调节。

(六) 5100 ICP-OES 原子发射光谱仪

(1) 性能与结构 5100 ICP-OES 采用同步垂直双向观测（SVDV）的智能光谱组合技术，可在一次读数中获得等离子体的水平和垂直观测结果，减少测量次数和氩气消耗。该仪器采用的 VistaChip Ⅱ 检测器是连续波长范围、零气体消耗的高速电荷耦合检测器（CCD），可快速预热，具有高通量、高灵敏度和宽的动态范围。冷锥接口（CCI）通过从轴向光路中去除冷等离子体尾焰来减少自吸收和重组干扰。固态射频系统提供可靠、稳定且无需维护的等离子体，可实现长期的分析稳定性。5100 电感耦合等离子原子发射光谱仪如图 2-6 所示。

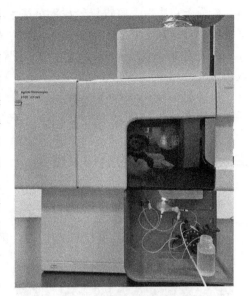

图 2-6 5100 电感耦合等离子原子发射光谱仪（彩图）

(2) 仪器操作步骤

① 开机预热

a. 确认有足够的氩气用于连续工作。

b. 确认废液收集桶有足够的空间用于收集废液。

c. 打开稳压电源开关，确保电源稳定。

d. 打开氩气瓶，并调节分压在 0.60～0.65MPa。保证仪器驱气 1h 以上。

e. 打开计算机。

f. 打开主机电源，仪器开始预热。

g. 待仪器自检完成后，双击"iTEVA"图标，启动 iTEVA 软件，进入操作软件主界面，仪器开始初始化。

② 编辑分析方法

a. 选择被测元素及其分析谱线。

b. 设置参数。

c. 设置工作曲线参数。

③ 点燃 ICP 炬

a. 再次确认氩气储量和压力，并确保驱气时间大于 1h，以防止 CID 检测器结霜，造成 CID 检测器损坏。

b. 光室温度稳定在（38±0.29）℃。CID 温度小于−40℃。

c. 检查并确认进样系统（炬管、雾化室、雾化器、泵管等）是否正确安装。

d. 夹好蠕动泵夹，把试样管放入蒸馏水中。

e. 开启通风。

f. 开启循环冷却水。

g. 打开 iTEVA 软件中"等离子状态"对话框，查看联锁保护是否正常，若有红灯警示，需做相应检查，若一切正常点击"等离子体开启"，点燃 ICP 炬。

h. 待等离子体稳定 15min 后，即可开始测定试样。

④ 绘制标准曲线并分析试样。

⑤ 关闭 ICP 炬

a. 分析完毕后，将进样管放入蒸馏水中冲洗进样系统 10min。

b. 在"等离子状态"对话框，点击"等离子关闭"，关闭 ICP 炬。

c. 关闭 ICP 炬 5~10min 后，关闭循环水，松开泵夹及泵管，将进样管从蒸馏水中取出。

d. 关闭排风。

e. 待 CID 温度升至 20℃以上时，驱气 20min 后，关闭氩气。

二、电化学仪器的结构及使用

（一）pHS-3C 型酸度计

（1）性能与结构　pHS-3C 型酸度计采用了高输入阻抗的运算放大器 AD515J 担任阻抗变换，由三位半数字电压表显示直流毫伏电势和 pH 值。仪器主要部分采用集成电路，线路简单、可靠性高。pHS-3C 型酸度计如图 2-7 所示。

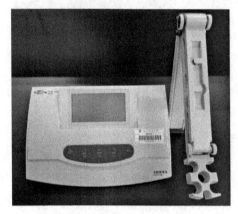

图 2-7　pHS-3C 型酸度计（彩图）

（2）溶液 pH 测定操作

① 安装电极。将 pH 玻璃复合电极插入电位计插孔，打开仪器开关，预热。

② 校正

a. 拔出 pH 玻璃复合电极保护帽，将电极用去离子水洗净，滤纸吸干。

b. 电极浸入 pH = 6.86 的标准溶液中，打开搅拌开关，用温度计测出溶液温度值，在仪器上设置此温度。

c. 示值稳定后，按"定位"键，显示屏显示"std yes"，按"确认"键，再次按"确认"键，仪器识别当前温度下标准溶液的 pH 值。

d. 拔出电极洗净，滤纸吸干后，浸入第二种标准溶液中（根据测试溶液的酸碱情况选择），按"斜率"键，按"确认"键，仪器识别当前温度下该溶液的 pH 值。

e. 经过标定的仪器可测量未知溶液 pH 值，不得再按"斜率"键和"定位"键。

③ 试液的 pH 测定。当被测溶液与标定溶液温度相同时，用去离子水清洗电极，滤纸吸干，将电极插入未知试液中，待显示屏上数据稳定后读出溶液的 pH 值。

（3）测量电势的操作。将电极插入被测液体，切换 pH/mV 功能键至"mV"模式，读取其平衡电位值。

（4）注意事项

① 玻璃电极的插口必须保持清洁，不使用时应将电极插头插入，以防止灰尘及湿气浸入。

② 玻璃电极球泡有裂纹或老化，则应调换新的电极。新电极在使用前需要用蒸馏水浸泡 24h。

（二）KLT-1 型通用库仑仪

（1）性能与结构 KLT-1 型通用库仑仪根据库仑滴定的基本原理，采用电位法、电流法、等当点上升、等当点下降四种指示电极终点检测方式，自动控制滴定终点，以积分方式计算、显示滴定过程中所消耗的电量。仪器分析精度较高，可以应用于环境监测，石油化工等多个领域。KLT-1 型通用库仑仪面板功能如图 2-8 所示。

图 2-8　KLT-1 型通用库仑仪（彩图）

（2）仪器操作步骤

① 开机准备。将电极安装至库仑仪，开机预热 20min。

② 空白溶液预电解

a. 将电极放入电解池，接上工作电极、对电极、终点指示电极。

注意：红色的接工作电极，黑色的接对电极，其余的接终点指示电极。

b. 在电解池中加入空白溶液，打开搅拌开关。

c. 按下"电流""上升""启动"键，调节补偿电位。

d. 复原"极化电位"键，按下"电解"键，"工作/停止"开关置"工作"，开始电解。

e. 库仑读数不变时，此为空白溶液消耗的电量。

③ 标准溶液电解

a. 取适量标准溶液至电解池，按下"电解"键，"工作/停止"开关置"工作"，开始电解。

b. 至电量不变、电流指针偏转，读取并记录其消耗电量。

④ 样品电解

a. 取适量被测液体至电解池，按下"电解"键，"工作/停止"开关置"工作"，开始电解。

b. 至电量不变、电流指针偏转时，读取并记录其消耗电量。

⑤ 关机。拆卸电极，关机。

（3）注意事项

① 电解液要经常配制，保持新鲜。

② 实验前要把电解池和电极清洗干净。

③ 电极的极性切勿接错，若接错必须把电极清洗干净。

④ 测量完毕，释放仪器面板上的所有按键，用蒸馏水清洗电极和电解池。关闭电源，盖好仪器罩。

（三）CHI660E 电化学工作站

（1）性能与结构 CHI660E 电化学工作站（图 2-9）为通用电化学测量系统。仪器内含快速数字信号发生器，用于高频交流阻抗测量的直接数字信号合成器。该仪器拥有双通道高速数据采集系统，电位电流信号滤波器，多级信号增益，iR 降（电压降）补偿电路，以及

图 2-9　CHI660E 电化学工作站（彩图）

恒电位仪/恒电流仪。电位范围为±10V，电流范围为±250mA。电流测量下限低于 10pA，可直接用于超微电极上的稳态电流测量。

CHI660E 系列仪器集成了几乎所有常用的电化学测量技术。为了满足不同的应用需要以及经费条件，CHI660E 系列分成多种型号。不同的型号具有不同的电化学测量技术和功能，但基本的硬件参数指标和软件性能是相同的。

（2）仪器操作步骤

① 打开仪器后部的开关。

② 将所需要检测的体系（一般为某物质的溶液）放置在烧杯或其他适合的容器中，将所需要采用的电极放置在溶液内。

③ 电极一般采用三电极系统，分别为工作电极、对电极、参比电极。接线如下：绿色夹头接工作电极，红色夹头接对电极，白色夹头接参比电极；在使用两电极系统的情况下接线方式如下：绿色夹头接工作电极，红色和白色夹头接另一电极。

④ 双击电脑上的 CHI660E 软件打开软件界面。

⑤ 点击软件中的"Setup"（设置）的菜单上找到"System"（系统）的命令，选择正确的接口，在此为 com2 口。

⑥ 点击软件中的"Setup"（设置）的菜单上找到"Hardware Test"（硬件测试）选项，进行系统测试，大约一分钟后屏幕上会显示硬件测试的结果。

⑦ 电化学技术测定：对于 CV（循环伏安法）扫描，在 CHI660E 软件界面打开"Technique"，点击"Cyclic voltammetry"，设定所需电势范围和扫描速度，点击"运行"即可。其他技术同 CV 进行。

⑧ 数据保存：点击"保存"按钮可保存原始图，点击"Data list"也可拷贝保存数据。

（3）注意事项

① 检测过程中不应出现电流溢出（overflow）现象，当软件显示电流过大的时候应及时停止实验，关闭仪器，检测电极系统之间是否有短路现象。

② 严禁将溶液等放置在仪器上方以防将溶液溅入仪器内部导致主板损毁。

③ 仪器应避免强烈振动或撞击。

（四）Titrando 888 型自动电位滴定仪

（1）性能与结构　Titrando 888 型自动电位滴定仪（图 2-10）具有滴定和加液功能，内置 EEPROM 智能芯片，能自动识别滴定管体积、滴定剂名称、滴定剂浓度和滴定度。

（2）仪器操作步骤

① 开机。将仪器主机电源和电脑连接线连接

图 2-10　Titrando 888 型自动
电位滴定仪（彩图）

好，双击桌面上的 Tiamo 图标，打开软件。

② 确认配置信息

a. 单击左侧下方按钮切换到配置界面，此界面分为 4 个子窗口，分别为设备、滴定剂/溶液、传感器、公共变量。

b. 检查设备中的仪器连接是否正常。

c. 检查溶液是否正常，在滴定剂/溶液窗口中，所需试剂一行，加液设备栏应显示该加液设备所在的 msb 接口。

d. 检查电极是否连接正常，在传感器窗口中，智能电极传感器类型为绿色字体，标明了电极类型，非智能电极传感器没有辨识标识。

③ 滴定剂准备

a. 清洗计量管和计量管单元的管路，并在计量管中排出气泡、使之充满试剂。

b. 点击屏幕左下方人工控制图标选择相应的加液设备。将滴定头放入废液杯中，点击"准备"，点击"开始"按键，将执行准备过程。

④ 测定

a. 单击"工作平台"按钮，切换到工作平台界面。

b. 在右上角的执行子窗口中，单击"方法"右侧的下拉箭头，在出现的方法中选择本次实验要使用的方法。此时您选中的方法会出现在左侧的"方法"子窗口中。

c. 准备好样品烧杯，放入搅拌器，放置在滴定台上，将电极和滴定头下移，使二者均浸入液面以下。注意：调整电极高度使之与搅拌器保持一定距离，防止电极受损。

d. 输入样品编号，样品量等信息，单击"开始"，实验开始，等待实验完成。

⑤ 数据查看。切换到数据库界面，单击"文件""开始"，在出现的对话框中选择对应的数据库，单击"打开"，即可在选定的数据库中查看您所需要的数据信息。

三、色谱分析仪器的结构及使用

(一) 7820A 气相色谱仪

(1) 性能与结构　气相色谱仪由气路系统、进样系统、分离系统、温控系统和检测记录系统组成。载气由高压钢瓶中流出，经减压阀降压到所需压力后，通过净化干燥管使载气净化，再流经稳压阀和转子流量计。随后，载气以稳定的压力、恒定的速度流经汽化室，与汽化的试样混合后，将试样气体带入色谱柱中进行分离。分离后的各组分随着载气先后流入检测器，然后载气放空。检测器将物质的浓度或质量的变化转变为一定的电信号，经放大后在记录仪上记录下来，就得到色谱流出曲线。7820A 气相色谱仪如图 2-11 所示。

根据色谱流出曲线上得到的每个峰的保留时间，可以进行定性分析，根据峰面积或峰高的大小，可以进行定量分析。

图 2-11　7820A 气相色谱仪（彩图）

(2) 仪器操作步骤

① 开机

a. 打开计算机电源和 GC 开关。

b. 打开氮气总阀，空气（全自动空气源）、氢气发生器开关。

c. 打开控制面板。

d. 设置参数，等待仪器检测器、柱子升温，仪器显示"就绪"。

② 测试

a. 点击"控制""单次运行"，编辑样品 ID，数据文件名，样品瓶号。

b. 打开已编辑好方法，保存路径，注册完成后点击"开始"。

c. 手动注入样品后，按下仪器面板上的"Start"按键，开始样品数据采集。

d. 单击"文件"，打开已完成样品数据谱图，点击"方法""积分事件"，进行谱图编辑、数据处理，保存已处理好数据。

③ 关机

a. 结束实验，编辑关机方法，降柱温到 50℃ 左右。

b. 待仪器温度降下后，关闭 GC 系统，退出工作站，关闭仪器主机电源，关闭氮气、氢气、空气。

(3) 注意事项

① H_2 比较危险，一定要经常检漏，不用时要立即关上。

② 柱子要老化后再接上检测器，以免流失造成喷嘴堵塞。

③ 进样口进样垫和进样口内的玻璃衬管要定期更换，不用的进样口和检测器要用堵头堵好。

(二) LC-20AD 高效液相色谱仪

(1) 仪器结构和工作原理　高效液相色谱仪主要包括高压输液系统、进样系统、分离系统、检测系统四个主要部分。LC-20AD 系统配备两个溶剂泵、一个内置脱气机、一个柱温箱、一个自动进样器，并由 LabSolutions 工作站软件控制（图 2-12）。

(2) 仪器操作步骤

① 开机。依次开启输液泵电源、真空脱气机、系统控制器、检测器系统、自动进样器、柱温箱的电源开关。打开电脑，最后点击"Lab Solutions"快捷键打开工作站界面。

② 分析方法编辑

a. 首先打开 A、B 泵上的排空阀（"开始"方向旋转 180°）。点击自动进样器面板上"Purge"（"排空"）键，自动排空 25min。

图 2-12　LC-20AD 高效液相色谱仪（彩图）

b. 点击仪器参数视图中"常规"按钮，设置测定时间、检测器波长、柱温箱温度、流速。保存方法文件，并单击"下载"将设置传送到仪器。

c. 点击"开启系统"键，此时泵开始工作，柱温开启。只有当仪器显示为"就绪"时，方可进样。

③ 进样

a.当基线及压力平稳后，准备进样。

b.选择"单次分析"，输入数据文件名称及保存路径，选定样品瓶号，输入进样量，点击确定后开始进样。

c.完成分析后，清洗色谱柱，若所用流动相含盐，首先需采用95%去离子水清洗20~30min；然后用纯甲醇冲洗30min即可。

④ 关机。清洗完成后，先关闭泵及柱温箱的加热，关闭各组件电源，关闭计算机。

（3）注意事项

① 高压恒流泵的密封圈是最易磨损的部件，密封圈的损坏可引起系统的许多故障，要注意保养和定期更换。

② 开始进样前，泵和检测器必须排空。

③ 必须使用 HPLC 级或相当于该级别的流动相，并先经 $0.45\mu m$ 薄膜过滤。过滤后的流动相必须经过充分脱气，以除去其中溶解的氧气等。如不脱气易产生气泡、基线噪声增加、灵敏度下降。

（三）BECKMAN P/ACE MDQ 毛细管电泳仪

（1）性能与结构 毛细管电泳，也称为高效毛细管电泳，是一类以毛细管为分离通道、以高压直流电场为驱动力的新型液相分离技术。BECKMAN P/ACE MDQ 系统的基本结构包括进样、填灌/清洗、电流回路、毛细管/温度控制、检测/记录/数据处理等部分（图2-13）。

（2）仪器操作步骤

① 开机。打开毛细管电泳仪开关，打开计算机。点击桌面 32 Karat 操作软件图标，点击 DAD（二极管阵列检测器）图标，进入毛细管电泳仪控制界面。

② 放样。将分别装有 0.1mol/L 盐酸、1mol/L 氢氧化钠、运行缓冲液 A、重蒸水依次放入左边缓冲液托盘 (inlet)，并记录对应的位置。然后将

图 2-13　BECKMAN P/ACE MDQ
高效毛细管电泳仪（彩图）

装有运行缓冲液 A 及空的缓冲液瓶放入右边缓冲液托盘 (outlet)，记录对应的位置。将装有待检测样品的缓冲液瓶放入左侧试样托盘，记录对应的位置。检查卡盘和试样托盘是否正确安装。关好托盘盖，注意图像屏幕上是否显示卡盘和托盘盖已安装好，此时应能听到制冷剂开始循环的声音。

③ 冲洗毛细管。在控制屏幕上点击压力区域，出现对话框；设置 Pressure（压力）、Duration（时长）、Direction（方向）、Pressure Type（压力类型）、Tray Positions（盘位置）等参数；点击"OK"，瓶子移到指定的位置，开始冲洗；冲洗完成后，毛细管中应充满运行缓冲液。

④ 方法编辑。进入 32 Karat 主窗口，选择"Open Offline"（离线打开）。选择"File/

Method/New"（文件/方法/新建），选择"Method/Instrument Setup"（方法/仪器设置），进入方法的仪器控制和数据采集模块；选择"Initial Condition（初始条件）"选项卡，在对话框中输入仪器运行时的参数。

⑤ 建立序列。从仪器窗口选择"File/Sequence/New"（文件/序列/新建），打开序列向导，按要求选择；点击"Finish"（完成），出现新建的序列表。

⑥ 系统运行。在系统运行前，检查仪器的状态。从菜单选择"Control/Single Run"（控制/单个样品分析），在仪器窗口的工具条上点击绿色的双箭头，打开运行序列对话框。

⑦ 关机。关闭氘灯，点击"Load"（装液），使托盘回到原始位置；打开托盘盖，待冷凝液回流后关闭控制界面；关闭毛细管电泳仪开关，关闭计算机，切断电源。

（3）注意事项

① 运行同一缓冲液时只需用该缓冲液冲洗 3min，否则需用高纯水冲洗后再用缓冲液冲洗。

② 关机淋洗后，进出口均用蒸馏水封住；长期不使用仪器，需吹干毛细管后，用空瓶封住。

③ 仪器运行过程中产生高压，严禁打开托盘盖。

（四）Metrohm 861 型离子色谱仪

（1）性能与结构 Metrohm 861 型离子色谱仪（图 2-14）是一种双抑制型离子色谱仪，

自带电导检测器。也可外接紫外可见检测器（UV/Vis）、二极管阵列检测器（DAD）、伏安检测器（VA）和脉冲安培检测器（PAD），还可以和等离子体光谱/质谱（ICP-AES/MS）联用。采用不同的离子交换柱，可以对试样中的阳离子或阴离子进行分离，并可根据离子色谱峰的峰高或峰面积进行定量分析。

（2）仪器操作步骤

① 开启系统。双击桌面离子色谱软件图标，进入操作软件。

② 预热准备。打开系统窗口，在系统窗口中点击"系统、更改"，更改系统为"阴离子系统平衡"。点击"控制、开始测定"（确认每过 10min 抑制器切换后有一水负峰）。预热 30～60min 直至基线平衡，点击"控制、停止测定"。

③ 试样准备。标样可采用一次性注射器直接进样，试样需用 $0.45\mu m$ 孔径过滤膜过滤后进样，未知试样还需先稀释 100～1000 倍后再进样，确保浓度不会太高进而污染系统。

图 2-14 Metrohm 861 型离子色谱仪（彩图）

④ 测定。预热结束后，在系统窗口中点击"系统、更改"，更改系统为"阴离子试样分析"。点击"控制、开始测定"，在弹出的对话框中输入样品信息（试样为 0，标样为 1、2、3……）。点击"确定"，将试样通过注射器注入定量环（注意下一次进样前不要取下注射器）。若想更改采样时间，可点击"方法、属性"，输入采样时间。多个试样测定重复上述步骤。若中间有预计 1h 以上的休息时间，请将系统方法切换到"阴离子系统平衡"，否则抑制器会饱和。

⑤ 关闭系统。测定结束后，在系统窗口中点击"控制、关闭硬件"，关闭整个系统，最后关闭计算机和离子色谱仪电源。

(3) 注意事项

① 长时间不使用的试剂不得存放于仪器托盘中，特别是盐酸，有可能造成仪器部件的腐蚀和仪器内的湿度增加。

② 未涂层的毛细管长时间不用，要先用水清洗，再用空气吹干。

③ 长时间不使用，在停机之前必须使样品及缓冲溶液托盘处于"Load"（负载）状态。

（五）7890B-5977B GC-MSD 气质联用仪

(1) 性能与结构 Agilent 7890B 气相色谱（GC）系统具有准确的温度控制和准确的进样系统，以及高性能电子气路控制（EPC）模块，可获得更佳保留时间和峰面积重现性。Agilent 5977B MSD 高效离子源（HES）和 InertPlus Extractor EI 离子源可使由离子源体产生并传输至四极杆分析器的离子数量大大增加。该联用仪器（图 2-15）广泛应用于环境、化学、石油化工、食品、法医、制药和材料等领域的测试。

图 2-15　7890B-5977B GC-MSD
气质联用仪（彩图）

(2) 仪器操作步骤

① 开机

a. 打开载气钢瓶控制阀，设置分压阀压力至 0.5MPa。

b. 打开计算机，依次打开 7890B GC、5977B MSD 电源（若 MSD 真空腔内已无负压则应在打开 MSD 电源的同时用手向右侧推真空腔的侧板直至侧面板被紧固地吸牢），等待仪器自检完毕。

c. 在桌面双击 Instrument♯1 图标，进入 MSD 化学工作站。

② 检查真空状态。在"Instrument Control"界面下，单击"View"菜单，选择"Tune and Vacuum Control"，进入调谐与真空控制界面。在"Vacuum"菜单中选择"Vacuum Status"，观察真空泵运行状态。状态应显示涡轮泵转速（turbo pump speed）很快达到100%，否则，说明系统有漏气，应检查侧板是否压正、放空阀是否拧紧、柱子是否接好。

③ 检查水峰及空气峰。在"Instrument Control"界面下，单击"View"菜单，选择"Tune and Vacuum Control"，在"Parameter"下选择"Manul Tune"，检查水峰、空气峰是否符合要求。

④ 调谐（一般选择自动调谐）。单击"View"菜单，选择"Tune and Vacuum Control"，单击"Tune"，选择"Autotune"，进行自动调谐，调谐结果自动打印。

⑤ 配置编辑

a. 点击"Instrument"菜单，选择"GC Configuration"。

b. 在"Connection"页面下，输入 GC Name，如"GC 7890B"；可在"Notes"处输入"7890B"的配置，如"7890B GC with 5977B MSD"。点击"Get GC Configuration"获取

7890B 的配置。

c. 依次进行 ALS、模块、色谱柱等设定。

⑥ 方法编辑

a. "Method" 菜单中选择 "Edit Entire Method" 项，选中除 "Data Analysis（数据分析）" 外的三项，点击 "OK"。

b. 编辑关于该方法的注释，然后点击 "OK"。

c. 设置色谱参数如进样器、柱模式、阀模式、分流与不分流进样、柱温箱、ALS、时间表、信号及其他参数的设定。

d. 选择质谱扫描模式，并设置相关参数。

⑦ 数据采集与处理

a. 在 "Method" 菜单下点击 "Run Method" 运行方法。

b. 双击桌面上的 "Instrument ♯1 Data Analysis" 图标，打开 MSD 的 Data Analysis，进行数据分析与处理。

⑧ 关机

a. 在 "Instrument Control" 界面下，单击 "View" 菜单，选择 "Tune and Vacuum Control"，选择 "Vent"，在跳出画面中点击 "OK" 进入放空程序。

b. 等到涡轮泵转速降至 0，同时离子源和四级杆温度降至 100℃以下，大概 40min 后退出工作站软件，并依次关闭 GC、MSD 电源，最后关闭载气。

（六）6230 TOF 液质联用仪

（1）性能与结构　安捷伦（Agilent）6230 飞行时间液质联用系统（TOF LC/MS）（图 2-16）用于分析、鉴定、表征和定量低分子量化合物和生物分子。该仪器集成三核技术创新即 TOF 技术，热聚焦技术和 MassHunter 工作站软件。6230 TOF 平台非常适合复杂样品的精确质量分析。

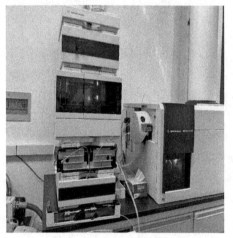

图 2-16　6230 飞行时间液质联用系统
（TOF LC/MS）（彩图）

（2）仪器操作步骤

① 打开氮气发生器的电源，待压力输出稳定后，调节输出压力为约 0.7 MPa（～110 psi）。

② 打开不间断电源，确认其处于正常状态。

③ 准备液相所需流动相、泵柱冲洗溶剂，检查管线连接状态，确认出真空泵和喷雾室的废气排到实验室外部。

④ 依次打开液相各模块电源及质谱电源，等待各模块自检完成。

⑤ 打开计算机及网络交换机电源。

⑥ 打开质谱诊断软件，双击 "Pump Down"（抽气）及 "HV condition"（高压条件），等待 TOF 高真空小于 $1.5×10^{-6}$ 及高压的设置条件完成。

⑦ 双击 "MassHunter" 采集软件图标，进入 "MassHunter" 工作站。

⑧ 进入调谐界面，进行质量轴校准。

⑨ 根据所测样品情况，建立液相色谱方法，根据样品结构选择正负离子模式及相应的质谱方法。创建并运行工作列表。

⑩ 序列运行结束，清洗管路，打开数据分析软件，进行数据分析。

⑪ 测样结束，停止运行。关闭采样软件，关闭液相各个模块，清洗离子源，冲洗雾化器组件。

⑫ 短时间内质谱仪在不运行样品的情况下处于待机状态即可。长期不用的情况下需要关机。

第三章　仪器分析实验

实验1　邻二氮菲分光光度法测定蜂蜜中铁含量
（见视频1）

实验目的

（1）了解分光光度计的结构和掌握分光光度计的正确使用方法。

（2）学习如何选择吸光光度分析的实验条件。

（3）学习吸收曲线、工作曲线的绘制及最大吸收波长的选择。

视频1

实验原理

吸光光度法是基于物质分子对光的选择性吸收而建立的分析方法。在吸光光度法分析中，如果待测组分本身有色，吸光能力较强，就可以直接进行测定。若待测组分本身无色或颜色较浅，则需先选择适当的试剂与试样中的待测组分反应使之生成吸光能力更强的有色化合物，然后再进行测定。将无色或浅色组分转变为有色化合物（络合物、离子或中性分子）的反应称为显色反应，所用试剂称为显色剂。

显色反应主要有氧化还原反应和络合反应两大类。在 pH＝2～9 的溶液中，邻二氮菲（Phen）与 Fe^{2+} 络合使颜色很浅的 Fe^{2+} 转变成稳定的邻二氮菲与 Fe^{2+} 的橙红色络合物，其稳定常数 $K_{稳}=10^{21.3}$，摩尔吸光系数 ε 为 $1.1\times10^4\,L/(mol \cdot cm)$，其反应式如下：

$$Fe^{2+}+3 \quad \longrightarrow \quad \left[\quad Fe \quad \right]_3^{2+}$$

该红色络合物的最大吸收峰在 510 nm 波长处。本方法的选择性很强，相当于含铁量 40 倍的 Sn^{2+}、Al^{3+}、Ca^{2+}、Mg^{2+}、Zn^{2+}、SiO_3^{2-}，20 倍的 Cr^{3+}、Mn^{2+}、V^{5+}、PO_4^{3-}，5 倍的 Co^{2+}、Cu^{2+} 等均不干扰测定。

蜂蜜是蜜蜂利用开花植物的花蜜制成的甜美液体。大约有 320 种不同的蜂蜜品种，它们的颜色、气味和风味各不相同。蜂蜜的主要成分是糖，还含有氨基酸、维生素、矿物质、铁、锌和抗氧化剂等。本实验采用邻二氮菲分光光度法测定蜂蜜中的铁含量。

仪器与试剂

(1) 仪器。721（或 722）分光光度计，10mL 吸量管，50mL 容量瓶，1000mL 容量瓶，1cm 比色皿，瓷坩埚，电炉，马弗炉。

(2) 试剂

① 0.0001mol/L 铁标准溶液：准确称取 0.0482g $NH_4Fe(SO_4)_2 \cdot 6H_2O$ 于烧杯中，用 30mL 2mol/L HCl 溶解，然后转移至 1000mL 容量瓶中，用水稀释至刻度，摇匀（供测摩尔比用）。② 0.1mg/L 的铁标准溶液：准确称取 0.8634 g 的 $NH_4Fe(SO_4)_2 \cdot 6H_2O$，置于烧杯中，加入 20mL（1∶1）HCl 和少量水，溶解后，定量地转移至 1 L 容量瓶中，以水稀释至刻度，摇匀。③ 邻二氮菲（1.5g/L 新配制的水溶液）。④ 盐酸羟胺 100g/L 水溶液（临用时配制）。⑤ 醋酸钠溶液（1mol/L）。⑥ NaOH 溶液（0.1mol/L）。⑦ HCl 溶液（1∶1）。⑧ 蜂蜜。

实验步骤

(1) 实验条件优化

① 吸收曲线的制作和测量波长的选择。用吸量管吸取 0mL、1.0mL 铁标准溶液，分别注入两个 50mL 容量瓶（或比色管）中，各加入 1mL 盐酸羟胺溶液，2mL 邻二氮菲，5mL NaAc，用水稀释至刻度，摇匀。放置 10min 后，用 1cm 比色皿，以空白试剂（即 0.0mL 铁标准溶液）为参比溶液，在 400～600nm 之间，每隔 10nm 测一次吸光度，在最大吸收峰附近，每隔 5nm 测定一次吸光度。在坐标纸上，以波长 λ 为横坐标，吸光度 A 为纵坐标，绘制 A 和 λ 关系的吸收曲线。从吸收曲线上选择测定 Fe 的适宜波长，一般选用最大吸收波长 λ_{max}。

② 溶液酸度的选择。取 7 个 50mL 容量瓶（或比色管），各加 1mL 的标准溶液，1mL 盐酸羟胺，2mL Phen，摇匀。然后，用滴定管分别加入 0mL、2.0mL、5.0mL、10.0mL、15.0mL、20.0mL、30.0mL 浓度为 0.10mol/L 的 NaOH 溶液，用水稀释至刻度，摇匀，放置 10min。用 1cm 比色皿，以蒸馏水为参比溶液，在选择的波长下测定各溶液的吸光度。同时，用 pH 计测量各溶液的 pH 值。以 pH 值为横坐标，吸光度 A 为纵坐标，绘制 A 与 pH 值关系的酸度影响曲线，得出测定铁的适宜酸度范围。

③ 显色剂用量的选择。取 7 个 50mL 容量瓶（或比色管），各加 1mL 铁标准溶液，1mL 盐酸羟胺，摇匀。再分别加入 0.1mL、0.3mL、0.5mL、0.8mL、1.0mL、2.0mL、4.0mL Phen 和 5.0mL NaAc 溶液，用蒸馏水稀释至刻度，摇匀，放置 10min。用 1cm 比色皿，以蒸馏水为参比溶液，在选择的波长下测定各溶液的吸光度。以所取 Phen 溶液体积 V 为横坐标，吸光度 A 为纵坐标，绘制 A 与 V 的显色剂用量影响曲线。得出测定铁时显色剂的最适宜用量。

④ 显色时间。在一个 50mL 容量瓶（或比色管）中，加入 1mL 铁标准溶液，1mL 盐酸羟胺溶液，摇匀。再加入 2mL Phen，5mL NaAc，以水稀释至刻度，摇匀。立即用 1cm 比色皿，以蒸馏水为参比溶液，在选择的波长下测量吸光度。然后依次测量放置 5min、10min、30min、60min、120min 后的吸光度。以时间 t 为横坐标，吸光度 A 为纵坐标，绘制 A 与 t 的显色时间影响曲线。得出铁与邻二氮菲显色反应完全所需要的适宜时间。

⑤ 邻二氮菲与铁的摩尔比的测定。取 50mL 容量瓶 8 个，吸取 0.0001mol/L 铁标准溶

液 10mL 于各容量瓶中，各加 1mL 10％盐酸羟胺溶液，5mL 1mol/L NaAc 溶液。然后依次加 0.02％邻二氮菲溶液（约为 1×10^{-10} mol/L）0.5mL、1.0mL、2.0mL、2.5mL、3.0mL、3.5mL、4.0mL、5.0mL，以水稀释至刻度，摇匀。然后在 510nm 的波长下，用 2cm 比色皿，以蒸馏水为空白液，测定各溶液的吸光度。最后以邻二氮菲与铁的浓度比 c_R/c_{Fe} 为横坐标，对吸光度作图，根据曲线上前后两部分延长线的交点位置确定 Fe^{2+} 与邻二氮菲反应的络合比。

（2）铁含量的测定

① 标准曲线的制作。用移液管吸取 100g/mL 铁标准溶液 10mL 于 100mL 容量瓶中，加入 2mL 2mol/L 的 HCl，用水稀释至刻度，摇匀。此溶液每毫升含 Fe^{3+} 10μg。在 6 个 50mL 容量瓶（或比色管）中，用吸量管分别加入 0mL、2.0mL、4.0mL、6.0mL、8.0mL、10.0mL 10μg/mL 铁标准溶液，分别加入 1mL 盐酸羟胺、2mL Phen、5mL NaAc 溶液，每加一种试剂后摇匀。然后，用水稀释至刻度，摇匀后放置 10min。用 1cm 比色皿，以空白试剂为参比溶液（即 0.0mL 铁标准溶液），在所选择的波长下，测量各溶液的吸光度。以含铁量为横坐标，吸光度 A 为纵坐标，绘制标准曲线。用绘制的标准曲线，重新查出相应铁浓度的吸光度，计算 Fe^{2+}-Phen 络合物的摩尔吸光系数 ε。

② 蜂蜜中铁含量的测定。准确移取蜂蜜 3.5～4.0g 于干净瓷坩埚中，在电炉上加热至不冒烟后，放入马弗炉中 850℃灰化 1.5h，冷至室温后，加入 2mL（1∶1）HCl，加热煮沸，加 5～10mL 蒸馏水，定量转移至 50mL 容量瓶后，加盐酸羟胺 2mL，Phen 2mL，1mol/L NaAc 10mL，加水稀释至刻度，摇匀。测量其吸光度 A。根据标准曲线求出试样中铁的含量（g/mL）。

数据处理

绘制各种条件试验曲线、标准曲线以及计算蜂蜜中铁的含量。

思考题

（1）本实验为什么要选择酸度、显色剂用量和有色溶液的稳定性作为条件实验的参数？

（2）吸收曲线与标准曲线有何区别？各有何实际意义？

（3）本实验中盐酸羟胺、醋酸钠的作用各是什么？

（4）怎样用吸光光度法测定试样中的全铁（总铁）和亚铁的含量？试拟出简单步骤。

（5）制作标准曲线和进行其他条件实验时，加入试剂的顺序能否任意改变？为什么？

实验 2 食品中 NO_2^- 含量的测定

实验目的

（1）进一步掌握可见分光光度计的操作原理。

（2）掌握食品中 NO_2^- 含量的测定方法。

实验原理

亚硝酸盐作为一种食品添加剂，能够保持腌肉制品等食物的色香味，并具有一定的防腐性。但同时也具有较强的致癌作用，过量食用会对人体产生危害。因此，食品加工中需严格控制亚硝酸盐的加入量。

在弱酸性溶液中，亚硝酸盐与对氨基苯磺酸发生重氮反应，生成的重氮化合物与盐酸乙二胺偶联成紫红色的偶氮染料，可用分光光度法测定，有关反应如下：

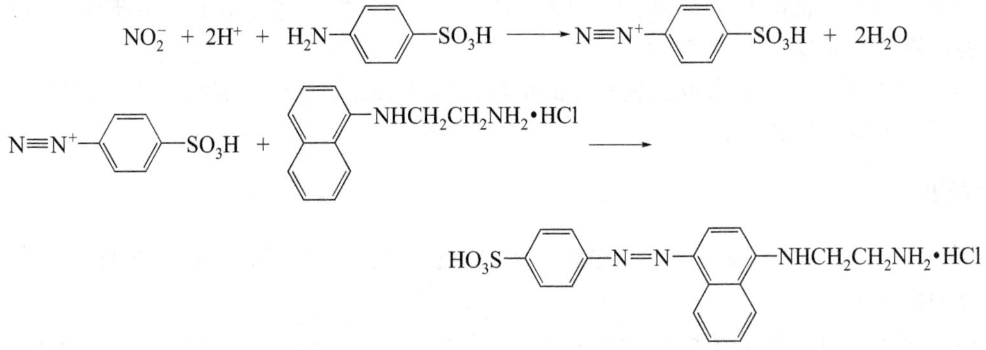

仪器与试剂

（1）**仪器**。721型分光光度计，小型多用食品粉碎机。

（2）**试剂**

①饱和硼砂溶液：称取 25g 硼砂（$Na_2B_4O_7 \cdot 10H_2O$）溶于 500mL 热水中。②1.0mol/L 硫酸锌溶液：称取 150g $ZnSO_4 \cdot 7H_2O$ 溶于 500mL 水中。③150g/L 亚铁氰化钾水溶液。④4g/L 对氨基苯磺酸溶液：称取 0.4g 对氨基苯磺酸溶于 200g/L 盐酸中配成 100mL 的溶液，避光保存。⑤2g/L 盐酸萘乙二胺溶液：称取 0.2g 盐酸萘乙二胺溶于 100mL 水中，避光保存。⑥$NaNO_2$ 标准溶液：准确称取 0.1000g 干燥 24h 的分析纯 $NaNO_2$，用水溶解后定量转入 500mL 容量瓶中，加水稀释至刻度并摇匀。临用时准确移取上述贮备液（0.2g/L）5.0mL 于 100mL 容量瓶中，加水稀释至刻度，摇匀，作为操作液（$1\mu g/mL$）。⑦活性炭。

实验步骤

（1）**试样预处理**

① 肉制品（如香肠，火腿）。称取 5g 经绞碎均匀的肉制品试样置于 50mL 烧杯中，加入 12.5mL 硼砂饱和溶液搅拌均匀，然后用 150～200mL 70℃以上的热水将烧杯中的试样全部洗入 250mL 容量瓶中，并置于沸水浴中加热 15min。取出，轻轻摇动，滴加 2.5mL $ZnSO_4$ 溶液以沉淀蛋白质。冷却至室温后，加水稀释至刻度，摇匀。放置 10min，撇去上层脂肪，清液用滤纸或脱脂棉过滤弃去最初 10mL 滤液，测定用滤液应为无色透明。

② 水果、蔬菜罐头。开启罐头，将内容物全部转至搪瓷盘中，切成小块混合均匀，用四分法取出 200g。将试样置于食品粉碎机的大杯内，加水 200mL，捣碎成匀浆后全部移入 500mL 烧杯中备用。称取匀浆 40g 于 50mL 烧杯中，用 150mL 70℃以上的热水分 4～5 次将其全部洗入 250mL 容量瓶中，加入 6mL 饱和硼砂溶液并摇匀。再加入 2g 经处理的活性炭，摇匀。然后加入 2mL $ZnSO_4$ 溶液和 2mL 亚铁氰化钾溶液，振摇 3～5min，再加水稀释至刻度，摇匀后用滤纸过滤，弃去最初的 10mL 滤液，承接其后滤液 50mL 左右用于测定。

（2）样品测定

① 标准曲线的绘制。准确移取 $NaNO_2$ 操作液（10g/mL）0mL、0.40mL、0.80mL、1.20mL、1.60mL、2.00mL 分别置于 50mL 的容量瓶中，各加水 30mL，然后分别加入 2mL 对氨基苯磺酸溶液，摇匀。静置 3min 后，再分别加入 1mL 盐酸乙二胺溶液，加水稀释至刻度，摇匀。放置 15min，用 1cm 吸收池，以空白试剂为参比溶液，于波长 540nm 处测定各试液的吸光度。

② 试样的测定。准确移取经过处理的试样滤液 40mL 于 50mL 容量瓶中，定容后重复（2）①操作步骤测得试样的吸光度。

数据处理

（1）以 $NaNO_2$ 溶液的浓度为横坐标，相应的吸光度为纵坐标，绘制标准曲线，得到一元线性回归方程。

（2）将样品的吸光度值带入回归方程，计算 $NaNO_2$ 的浓度。最后计算试样中 $NaNO_2$ 的质量分数（以 mg/kg 表示）。

注意事项

（1）亚硝酸盐容易氧化为硝酸盐，处理试样时加热的时间和温度均要注意控制，另外，配制的标准贮备液不宜久存。

（2）本法测量中不包括试样中硝酸盐的含量。

思考题

（1）查阅有关资料，了解金华火腿制作过程中食品添加剂的种类和用量。

（2）收集滤液时，为什么要弃去最初的 10mL 滤液？

实验 3　紫外吸收光谱法测定蒽醌的含量和摩尔吸光系数（见视频 2）

实验目的

（1）掌握紫外可见分光光度计的使用方法。

（2）掌握定量测定蒽醌的方法。

视频 2

实验原理

具有不饱和结构的有机化合物，如芳香族化合物，在紫外区（200~400nm）有特征的吸收，为这种有机化合物的鉴定提供了条件。

蒽醌在波长 251nm 处（$\lambda_{max}=251nm$）有强烈吸收峰 [$\varepsilon = 4.6 \times 10^4$ L/(mol·cm)]，在波长 323nm 处有中等强度的吸收峰 [$\varepsilon = 4.7 \times 10^3$ L/(mol·cm)]。为避免邻苯二甲酸酐在 251nm 附近吸收的干扰，实验选用 323nm 波长为定量测定蒽醌的入射光波长（图 3-1）。

光的吸收程度 A 与蒽醌的浓度 c 成正比（$A = \varepsilon b c$），采用标准曲线法即可求得样品的含

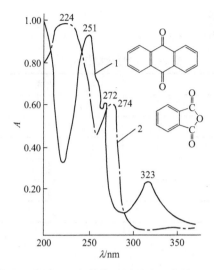

图 3-1　蒽醌（曲线 1）和邻苯二甲酸酐（曲线 2）吸收图谱

量。摩尔吸光系数是衡量物质在特定波长下对光吸收程度的重要指标，可用求标准曲线斜率的方法求得。

仪器与试剂

（1）**仪器**。普析 T9 紫外可见分光光度计，1cm 带盖石英比色皿，容量瓶（10mL，100mL，1000mL）。

（2）**试剂**

①蒽醌（分析纯）。②甲醇。③蒽醌试液。④0.1600g/L 蒽醌贮备液：准确称取 0.1600g 蒽醌于 100mL 烧杯中，用甲醇溶解后，转移到 1000mL 容量瓶中，用甲醇稀释至刻度，摇匀。⑤0.0640g/L 蒽醌标准溶液：吸取 40mL 0.1600g/L 蒽醌贮备液于 100mL 容量瓶中，用甲醇稀释至刻度，摇匀备用。

实验步骤

（1）**蒽醌系列标准溶液的配制**。分别吸取 1.00mL、2.00mL、3.00mL、4.00mL、5.00mL 0.0640g/L 蒽醌标准溶液于 5 只 10mL 容量瓶中，用甲醇定容，摇匀。

（2）**吸收曲线的扫描和测量波长的选择**

① 将 2 个装有甲醇溶液的比色皿置于吸收池中，在 200～350nm 范围内，扫描基线。

② 将其中的一份甲醇溶液替换为 4 号蒽醌标准溶液，在 200～350nm 范围内，扫描蒽醌的紫外吸收光谱，找到最大吸收波长并确定定量入射光波长。

（3）**标准曲线的绘制**。以甲醇溶液为参比，由低浓度到高浓度依次测定蒽醌系列标准溶液在 323nm 处的吸光度值。

（4）**试样中蒽醌的测定**。准确吸取 5mL 蒽醌试液，用甲醇定容至 10mL。与标准曲线相同条件下测量其吸光度，计算试样中蒽醌的含量（g/L）。

数据处理

（1）记录各实验数据于表 3-1 中，以蒽醌浓度为横坐标，吸光度为纵坐标，绘制标准曲线。

(2) 计算样品中蒽醌浓度。

(3) 计算蒽醌的摩尔吸光系数 ε。

表 3-1　实验数据记录及处理

λ/nm				参比溶液		
项目	标准溶液					样品
序号	1	2	3	4	5	6
V/mL	1.00	2.00	3.00	4.00	5.00	5.00
$c/(g/L)$						
A						
$\varepsilon/[L/(mol \cdot cm)]$						

思考题

(1) 为什么选用 323nm 而不选用 251nm 波长作为蒽醌定量分析的测定波长？

(2) 什么是摩尔吸光系数？它和分析物质的浓度有关吗？为什么？

实验 4　分光光度法同时测定铬和钴的含量

实验目的

(1) 进一步熟悉紫外可见分光光度计的使用。

(2) 掌握混合物分析的方法。

实验原理

紫外可见吸收光谱法采用朗伯-比尔定律进行定量分析。进行两种或多种吸收物质的测定时，各组分对光的吸收是有加和性的。如果各组分之间不发生反应，则在波长 λ 处，总吸光度 (A) 为各物质的吸光度之和 [式(3-1) 式(3-2)]：

$$A_\lambda = A_{1\lambda} + A_{2\lambda} + A_{3\lambda} + \cdots \tag{3-1}$$

$$A_\lambda = \varepsilon_{1\lambda} bc_1 + \varepsilon_{2\lambda} bc_2 + \varepsilon_{3\lambda} bc_3 + \cdots \tag{3-2}$$

下标 1，2 和 3 分别指不同的吸收组分。ε_λ 为波长 λ 处时物质的摩尔吸收系数。本实验的目的是同时测定 $[Cr(H_2O)_6]^{3+}$ [六水合铬 (Ⅲ)] 和 $[Co(H_2O)_6]^{2+}$ [六水合钴 (Ⅱ)] 的浓度。对于这种双组分混合物，根据吸光度的加和性原理可得联立方程式(3-3) 和式(3-4)：

$$A_{\lambda 1} = \varepsilon_{Cr\lambda 1} bc_{Cr} + \varepsilon_{Co\lambda 1} bc_{Co} \tag{3-3}$$

$$A_{\lambda 2} = \varepsilon_{Cr\lambda 2} bc_{Cr} + \varepsilon_{Co\lambda 2} bc_{Co} \tag{3-4}$$

其中 $\varepsilon_{Cr\lambda 1}$、$\varepsilon_{Cr\lambda 2}$、$\varepsilon_{Co\lambda 1}$ 和 $\varepsilon_{Co\lambda 2}$ 为 Cr 和 Co 在 λ_1 和 λ_2 处的摩尔吸收系数，可分别通过一定浓度的纯 Cr 和 Co 溶液在 λ_1 和 λ_2 处测量计算而得。解上述方程组即可求得混合物中 c_{Co} 和 c_{Cr}。

在该实验中，波长的选择非常重要。最理想的情况是选择一种组分完全不吸收，而对另一种组分基本吸收的波长，这将有利于把联立方程问题简单化；通常情况下，可以选择一个波长处大部分是吸收 Co（Ⅱ），而在另一个波长处大部分是吸收 Cr（Ⅲ）。

仪器与试剂

(1) 仪器。普析 T9 紫外可见分光光度计，1cm 带盖石英比色皿。

(2) 试剂。0.01mol/L Cr(NO$_3$)$_3$·9H$_2$O，0.04mol/L Co(NO$_3$)$_2$·6H$_2$O，未知混合物。

实验步骤

(1) 标准溶液的配制

① 用移液管分别取 0.010mol/L Cr(NO$_3$)$_3$·9H$_2$O 溶液 1mL、2mL、3mL、4mL 于 4 个 100mL 容量瓶中，加水稀释至刻度，配制浓度分别为 0.0001mol/L、0.0002mol/L、0.0003mol/L 和 0.0004mol/L 的 Cr（Ⅲ）标准溶液。

② 用移液管分别取 0.040mol/L Co(NO$_3$)$_2$·6H$_2$O 溶液 1mL、2mL、3mL、4mL 于 4 个 100mL 容量瓶中，加水稀释至刻度，配制浓度分别为 0.0004mol/L、0.0008mol/L、0.0012mol/L 和 0.0016mol/L 的 Co（Ⅱ）标准溶液。

(2) λ_1 和 λ_2 的选择。在 350～650nm 范围内，绘制 0.0002mol/L 和 0.0008mol/L Cr（Ⅲ）和 Co（Ⅱ）标准溶液的吸收光谱图，根据波长选择原则确定波长 λ_1 和 λ_2。

(3) $\varepsilon_{Cr\lambda1}$、$\varepsilon_{Cr\lambda2}$、$\varepsilon_{Co\lambda1}$ 和 $\varepsilon_{Co\lambda2}$ 测定。测定 Cr（Ⅲ）和 Co（Ⅱ）标准溶液分别在 λ_1 和 λ_2 处的吸光度，记录数据。

(4) 未知样品测定。测量两种波长下未知混合物的吸光度。

数据处理

(1) 分别以 λ_1 和 λ_2 处两种标准溶液的吸光度值为纵坐标，标准溶液的浓度为横坐标，绘制标准曲线并得到回归方程，由直线的斜率求得 $\varepsilon_{Cr\lambda1}$、$\varepsilon_{Cr\lambda2}$、$\varepsilon_{Co\lambda1}$ 和 $\varepsilon_{Co\lambda2}$。

(2) 计算未知混合溶液中 Cr（Ⅲ）和 Co（Ⅱ）的浓度。

思考题

(1) 简述紫外可见吸收光谱法测定双组分混合物的实验原理。

(2) 还可用什么仪器分析方法测定本实验中的混合物组分含量？

实验 5　标准加入分光光度法测定钢中锰含量

实验目的

(1) 理解光谱法测定钢中锰的原理。

(2) 掌握标准加入法的测定原理。

实验原理

钢是铁与少量过渡金属（如 Mn，Cr，Cu 等）的合金。钢中的过渡金属可在热的浓硝酸（4～5mol/L）中消解后，通过紫外可见吸收光谱法进行测定。通过消解，锰转化为无色

的 Mn^{2+}，氧化剂如高碘酸根离子进一步将 Mn^{2+} 氧化成深紫色的 MnO_4^-［式(3-5)］，得到的 MnO_4^- 可通过可见光谱定量法检测。其他金属可能会干扰 Mn 的分析，必须进行掩蔽或将其除去。钢的主要成分铁，可以通过添加磷酸（H_3PO_4）在水溶液中形成无色磷酸盐络合物进行掩蔽。其他金属如铬和铈在碘酸盐氧化过程中，可能形成干扰物质。为了考虑这些潜在的干扰离子的影响，本实验采用标准加入法进行定量分析。

$$2Mn^{2+} + 5IO_4^- + 3H_2O \Longleftrightarrow 5IO_3^- + 6H^+ + 2MnO_4^- \tag{3-5}$$

仪器与试剂

（1）**仪器**。722 型分光光度计，250mL 烧杯，10mL 吸量管，250mL 容量瓶，1L 容量瓶，电炉。

（2）**试剂**。钢样，4mol/L 硝酸（HNO_3），过硫酸铵［$(NH_4)_2S_2O_8$］，亚硫酸氢钠（$NaHSO_3$），纯锰金属，85％的磷酸，固体高碘酸钾（KIO_4）。

实验步骤

（1）**钢铁消化**。准确称量约 1g 钢样于 250mL 烧杯中。记录质量，精确到 0.1mg。向烧杯中加入 50mL 4mol/L HNO_3，小火煮沸几分钟直至样品溶解。在消化过程中，用表面皿覆盖烧杯，以防止材料因飞溅而损失。缓慢加入 1.0g 过硫酸铵［$(NH_4)_2S_2O_8$］并煮沸 10～15min 以氧化样品中可能存在的碳。如果此时样品呈粉红色或含有棕色沉淀，则加入约 0.1g 的亚硫酸氢钠（$NaHSO_3$）并继续加热 5min。溶液冷却至室温后，将溶液全部转入 250mL 容量瓶中，加水稀释定容、摇匀。

（2）**标准 Mn 溶液的配制**。准确称量约 100mg 的 Mn 金属，记录 Mn 的质量，精确到 0.1mg。将其溶解在 10mL 的 4mol/L HNO_3 中并煮沸几分钟以除去产生的氮氧化物，将溶液定量转移至 1L 容量瓶中，用蒸馏水稀释至刻度，摇匀。

（3）**标准加入法测量钢样中锰**。准确移取 4 份 20mL 的钢样溶液于 250mL 烧杯中，分别加入 5mL 85％的磷酸溶液。根据表 3-2 将标准溶液 Mn^{2+} 和固体 KIO_4 加入每份试样的烧杯中。将每种溶液煮沸 5min 并使其冷却至室温，将上述各溶液定量转移至 50mL 容量瓶中。使用无色空白溶液作为参比溶液，测量每种紫色溶液的吸光度。记录高锰酸根离子在 λ_{max} 处的吸光度。

表 3-2　标准加入法测量溶液中添加的样品量记录

容量瓶	钢样/mL	H_3PO_4/mL	标准溶液 Mn^{2+}/mL	KIO_4/g
1	0（空白）	5	0.00	0.0
2	20	5	0.00	0.4
3	20	5	2.50	0.4
4	20	5	5.00	0.4
5	20	5	10.00	0.4

数据处理

以吸光度为 y 轴，以添加的标准溶液 Mn^{2+} 的浓度为 x 轴，绘制标准曲线，其 x 轴截距为稀释后未知钢样中的 Mn 浓度。计算锰在钢样中的质量百分比（精确到 0.01％）。

注意事项

(1) 该方法适用于钢样品中锰浓度为 0.2% 至 0.5%。

(2) 如果溶液为粉红色或含有棕色的锰氧化物，则加入约为 0.1g 亚硫酸氢钠（NaHSO$_3$），再加热 5min。

思考题

(1) 简述标准加入法的测定原理。

(2) 讨论钢中锰元素含量还可用什么方法定量测定。

实验 6　硫酸阿托品片剂的染料比色法测定

实验目的

掌握酸性染料比色法的原理与方法。

实验原理

硫酸阿托品 $[(C_{17}H_{23}NO_3)_2 \cdot H_2SO_4 \cdot H_2O]$ 为无色结晶或白色结晶性粉末，无臭，在水中极易溶解，在乙醇中易溶，其结构式如图 3-2 所示。硫酸阿托品，是一种抗胆碱药，用于胃肠道、胆绞痛、散瞳检查验光、角膜炎、有机磷农药中毒、感染性休克等综合征的治疗。硫酸阿托品片规格为 0.3mg/片，含硫酸阿托品 $[(C_{17}H_{23}NO_3)_2 \cdot H_2SO_4 \cdot H_2O]$ 应为标示量的 90.0%～110.0%。

图 3-2　硫酸阿托品的结构式

硫酸阿托品片采用酸性染料比色法进行含量测定。在一定 pH 水溶液中，硫酸阿托品能与氢离子结合成阳离子（BH$^+$），而酸性染料溴甲酚绿可在同一 pH 条件下解离成阴离子（In$^-$）。此时阳离子和染料阴离子可以定量结合成黄色的有机配合物，即离子对（BH$^+$ · In$^-$），此离子对可以定量地被三氯甲烷萃取，并在 420nm 波长处测定其吸光度，与对照品比较，即可计算出硫酸阿托品片的标示量百分含量。

仪器与试剂

(1) 仪器。分液漏斗，紫外可见分光光度计，100mL 烧杯，50mL 容量瓶，25mL 容量瓶，100mL 容量瓶。

(2) 试剂。硫酸阿托品片（市售），硫酸阿托品对照品，三氯甲烷，溴甲酚绿，邻苯二甲酸氢钾，氢氧化钠。

溴甲酚绿溶液：称取溴甲酚绿 50mg，邻苯二甲酸氢钾 1.021g，加 0.2mol/L 氢氧化钠溶液 6.0mL 使其溶解，再用水稀释至 100mL，摇匀备用。

实验步骤

(1) 样品溶液的配制。取本品 20 片，精密称定，研细，精密称取适量（约相当于硫酸阿托品 2.5mg），置 100mL 烧杯中，加水振荡使硫酸阿托品溶解，将溶液全部转移至 50mL 容量瓶中，加水稀释至刻度。过滤，取过滤液，作为供试品溶液。

（2）对照品溶液的配制。精密称定硫酸阿托品对照品约 25mg，用水定容成 25mL 溶液。精密量取 5.00mL，置 100mL 容量瓶中，加水稀释至刻度，摇匀，作为对照品溶液。

（3）硫酸阿托品片的含量测定。精密量取供试品溶液和对照品溶液各 2.00mL，分别转入预先精密加入三氯甲烷 10mL 的分液漏斗中，各加溴甲酚绿溶液 2.0mL，振荡提取 2min 后，静置使分层，分别取澄清的三氯甲烷液，在波长 420nm 处分别测定其吸光度值。

数据处理

硫酸阿托品片的标示量计算公式如下：

$$标示量 = \frac{c_R \times \dfrac{A_x}{A_R} \times D \times \overline{W} \times 1.027}{W} \times 100\%$$

式中，A_x 和 A_R 分别为供试品和对照品的吸光度；c_R 为对照品的浓度；D 为稀释体积；W 为称样量；$\overline{W}$ 为平均片重；1.027 为分子量换算因数。

注意事项

（1）供试品与对照品应平行操作，尤其在萃取过程中，振荡的方法、次数、速度及力度等均应一致。

（2）分液漏斗必须洗净干燥，不得含水，微量的水会使三氯甲烷发生浑浊，影响比色。同时分液漏斗还应使用甘油淀粉做润滑剂，不能用凡士林，否则三氯甲烷有可能溶解凡士林，造成漏液。

（3）取三氯甲烷层时应"斩头去尾"，弃去少量初滤液，所取的三氯甲烷层过滤液必须澄清透明，不得混有水珠。

（4）酸性染料比色法中的 1.027 为含 1 分子结晶水硫酸阿托品样品与无水硫酸阿托品对照品的分子量换算因数。

$$\frac{(C_{17}H_{23}NO_3)_2 \cdot H_2SO_4 \cdot H_2O \text{ 的相对分子质量}}{(C_{17}H_{23}NO_3)_2 \cdot H_2SO_4 \text{ 的相对分子质量}} = \frac{694.84}{676.84} = 1.027$$

思考题

（1）为什么硫酸阿托品原料药的含量测定采用非水溶液滴定法，而硫酸阿托品片的含量测定则选用酸性染料比色法？

（2）影响酸性染料比色法的主要因素有哪些？最重要的条件是什么？

实验 7　紫外双波长法测定对氯苯酚存在时苯酚的含量

实验目的

（1）熟悉双波长分光光度法的测定原理。

（2）掌握选择测量波长和参比波长的方法。

实验原理

当 M，N 两组分处于同一溶液中，如果它们的紫外吸收光谱相互干扰，无法通过测量某一波长处的吸光度值来得到其中一种组分的含量时，可采用双波长法消除干扰，测定一种组分的含量。如图 3-3 所示，λ_2 是 M 组分的最大吸收波长，对于 M 物质，在两波长 λ_1 和 λ_2 处 ΔA_M 最大，对于 N 物质，在此两波长处 $\Delta A_N = 0$。如果分别在 λ_1 和 λ_2 处测量混合溶液的吸光度值，得 $\Delta A = (\varepsilon_M^{\lambda_2} - \varepsilon_M^{\lambda_1}) b c_M$ 由此可消除 N 物质的干扰，测定 M 的含量。

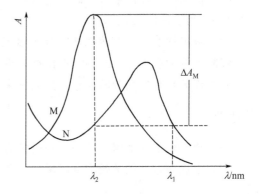

图 3-3 双波长法测定原理

双波长法所选择的波长，必须满足以下两个条件。首先，两波长处，干扰组分吸光度值相等。两波长处，待测组分的吸光度差值应足够大。

为了选择 λ_1 和 λ_2，应先在同一坐标体系里分别绘制两组分标准物质的紫外吸收光谱，再确定 λ_1 和 λ_2。其方法为在待测组分 M 的最大吸收峰处或其附近选择一测量波长 λ_2，由此作垂直于横轴的直线交干扰组分 N 的吸收光谱上某一点，再以此交点画一平行于横轴的水平线，与组分 N 的吸收光谱又产生一个或几个交点，交点处的波长即可作为参比波长 λ_1，当 λ_1 有几个位置可供选择时，所选择的 λ_1 应能使待测组分获得较大的吸光度差值。

本实验中，苯酚和对氯苯酚水溶液吸收光谱相互重叠，需用双波长法测定混合液中苯酚的含量。

仪器与试剂

（1）**仪器**。紫外可见分光光度计，石英比色皿 2 只（1cm）。

（2）**试剂**。250mg/L 苯酚和对氯苯酚储备液：分别准确称取 25.00mg 苯酚和 25.00mg 对氯苯酚，用无酚蒸馏水溶解，定量转移到 100mL 容量瓶中定容，摇匀。

实验步骤

（1）**苯酚和对氯苯酚水溶液吸收光谱的绘制**。分别将适量的储备液稀释 5 倍，配成 50.0mg/L 苯酚水溶液和 50.0mg/L 对氯苯酚水溶液，在 250～300nm 波长范围内，以无酚蒸馏水作参比，1cm 石英吸收池，用紫外可见分光光度计测绘它们各自的吸收光谱，并将两条吸收光谱绘制在同一坐标体系内，用作图法选择合适的 λ_1 和 λ_2。再用对氯苯酚水溶液复测两波长处其吸光度是否相等。

（2）**标准曲线的绘制**。分别移取 250mg/L 的苯酚水溶液 1.00mL、2.00mL、3.00mL、4.00mL、5.00mL 于 25mL 容量瓶中，用无酚蒸馏水定容摇匀。在所选择的测量波长 λ_2 及参比波长 λ_1 处，以无酚蒸馏水作参比液，用 1cm 石英吸收池，依次测定苯酚标准溶液的吸光度。

（3）**未知试样中苯酚的测定**。移取未知试样溶液 5.00mL 于 25mL 容量瓶中定容，在同样条件下测定试样溶液在 λ_2 及 λ_1 处的吸光度。

数据处理

（1）在同一坐标体系里绘制苯酚水溶液和对氯苯酚水溶液的吸收光谱，并选择合适的测量波长 λ_2 和参比波长 λ_1。

（2）求出标准系列溶液在两波长处吸光度的差值 $\Delta A_{\lambda_2 - \lambda_1}$，以 $\Delta A_{\lambda_2 - \lambda_1}$ 为纵坐标，苯酚水溶液的浓度为横坐标，绘制标准曲线，得到回归方程。将未知试样溶液的 $\Delta A_{\lambda_2 - \lambda_1}$ 值，代入回归方程计算相应的苯酚含量，然后求未知试样溶液中的苯酚浓度。

思考题

（1）双波长的仪器组成与单波长仪器有何异同点？

（2）本法所选择的波长应满足哪两个条件？

实验 8　荧光法定量测定维生素 B_2

实验目的

（1）熟悉荧光光谱仪的组成和使用。

（2）掌握荧光法测定维生素 B_2 含量的方法。

实验原理

荧光是光致发光，采用光电检测器，可以检测到荧光分子的两个过程，其一为激发过程，其二为分子的荧光发射过程。荧光的发射过程为电子从第一电子激发单重态的最低振动能级跃迁回到基态的过程。在稀溶液中，荧光发射强度与物质的浓度成正比，可进行物质的定量分析。

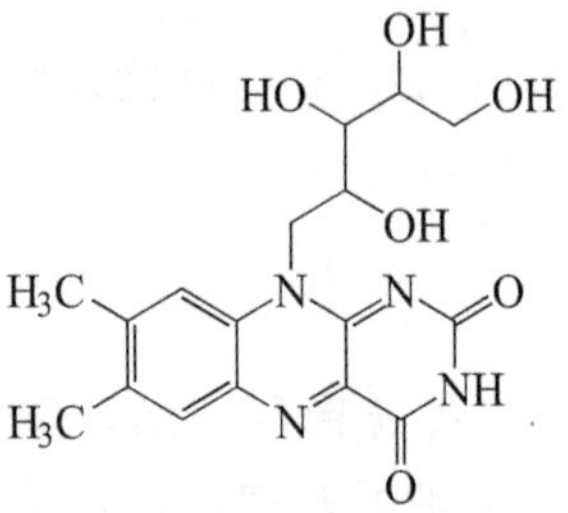

图 3-4　维生素 B_2 的结构式

一般来讲，具有强荧光性的物质，其分子结构往往具有以下特征：①具有大的共轭键体系；②具有刚性的平面结构；③共轭体系含有给电子取代集团；④其最低的电子激发单重态为 π，π^* 型。维生素 B_2 也叫核黄素，是橘黄色无臭的针状结晶，其结构式如图 3-4 所示。它具有刚性平面和良好的共轭结构，在弱酸性或中性水溶液中在 $430\sim440nm$ 的蓝光照射下，发出绿色荧光，其最大发射波长约为 525nm。在较低浓度时，物质的荧光发射强度 I 和它的浓度 c 成正比。本实验使用荧光分析法定量测定药物片剂中维生素 B_2 的含量。

仪器与试剂

（1）**仪器**。LS 55 荧光光谱仪，石英池，5mL 吸量管，50mL 容量瓶。

（2）**试剂**。维生素标准储备溶液（10.0mg/L），维生素 B_2 片剂。

实验步骤

（1）**系列标准溶液的配制**。分别准确移取 0.50mL、1.00mL、1.50mL、2.00mL 和

2.50mL 的维生素储备溶液（10.0mg/L）于 5 个 50mL 容量瓶中，用水稀释至刻度，摇匀。

（2）激发光谱和发射光谱的扫描。 根据 LS 55 荧光光度计的操作手册，设定激发波长为 425nm，在 500～600nm 的波长范围内扫描发射光谱，设定发射波长为 525nm，在 400～500nm 范围内扫描激发光谱。在激发和发射光谱上分别找出最大激发波长和最大发射波长。

（3）标准溶液的荧光测定。 在最佳激发波长下，扫描各标准溶液的荧光光谱，记录其在最大发射波长下的荧光强度。

（4）实际样品分析

① 精确称量和研磨 20 片维生素 B_2 片剂，称量研磨后的粉末 0.05～0.08g，加入 50mL 1% HAC 溶液，并在水浴中加热溶解 30min 后，将所得溶液定量转移到 250mL 容量瓶中，用蒸馏水稀释至刻度并摇匀。移取该溶液 1.00mL 至 50mL 容量瓶中，用蒸馏水稀释至刻度，摇匀。

② 在同样条件下测定药片溶液的荧光强度。

数据处理

（1）以各标准溶液的荧光强度值为纵坐标，其浓度为横坐标，得到标准曲线和线性拟合方程。

（2）根据试样的荧光强度，带入回归方程计算其浓度，并换算为原来片剂中维生素 B_2 的含量。

思考题

（1）说一说荧光光谱仪和分光光度计的异同点。

（2）如何绘制物质的荧光激发光谱和发射光谱？

实验 9　荧光分析法测定邻羟基苯甲酸和间羟基苯甲酸

实验目的

（1）掌握荧光分析法的基本原理和操作步骤。

（2）掌握用荧光分析法进行多组分含量测定的方法。

实验原理

邻羟基苯甲酸（亦称水杨酸）和间羟基苯甲酸分子组成相同，均含一个能发射荧光的苯环，但因其取代基位置的不同而具有不同的荧光性质。在 pH＝12 的碱性溶液中，二者在 310nm 附近紫外光的激发下均会发射荧光，在 pH＝5.5 的近中性溶液中，间羟基苯甲酸不发射荧光，邻羟基苯甲酸由于分子内形成氢键增加了分子刚性而发出较强的荧光，且荧光强度与 pH＝12 时相同。利用这一性质，可在 pH＝5.5 时测定二者混合物中邻羟基苯甲酸的含量，另取相同质量的混合物溶液，测定 pH＝12 的荧光强度，减去 pH＝5.5 时测得邻羟基苯甲酸的荧光强度，即可求出间羟基苯甲酸的含量。

仪器与试剂

（1）仪器。LS 55 荧光光谱仪，10mL 比色管，吸量管。

（2）**试剂**。60μg/mL 邻羟基苯甲酸标准溶液，60μg/mL 间羟基苯甲酸标准溶液，0.1mol/L NaOH 溶液。pH＝5.5 的醋酸-醋酸钠（HAc-NaAc）缓冲溶液：47g NaAc 和 6g 冰醋酸溶于水并稀释至 1L。

实验步骤

（1）标准系列溶液的配制

① 分别移取 0.40mL、0.80mL、1.20mL、1.60mL、2.00mL 邻羟基苯甲酸标准溶液于已编号的 10mL 比色管中，各加入 1.0mL pH＝5.5 的 HAc-NaAc 缓冲溶液，以蒸馏水稀释至刻度，摇匀。

② 分别移取 0.40mL、0.80mL、1.20mL、1.60mL、2.00mL 间羟基苯甲酸标准溶液于已编号的 10mL 比色管中，各加入 1.2mL 0.1mol/L 的 NaOH 水溶液，以蒸馏水稀释至刻度，摇匀。

③ 分别取未知溶液 2.00mL 于 2 个 10mL 比色管中，其中一份加入 1.00mL pH＝5.5 的 HAc-NaAc 缓冲溶液，另一份加入 1.20mL 0.1mol/L 的 NaOH 水溶液，以蒸馏水稀释至刻度，摇匀。

（2）扫描荧光激发光谱和发射光谱

固定发射波长为 400nm，以 250～350nm 进行激发波长扫描，获得邻羟基苯甲酸和间羟基苯甲酸溶液（第三份标准溶液）的激发光谱和荧光最大激发波长 λ_{ex}^{max}。固定激发波长 λ_{ex}，以 350～500nm 进行发射波长扫描，获得溶液的发射光谱和荧光最大发射波长 λ_{em}^{max}。

（3）荧光强度测定

根据上述激发光谱和发射光谱扫描结果，确定一组波长（λ_{ex} 和 λ_{em}）使之对两组分都有较高的灵敏度，并在此组波长下测定前述标准系列各溶液和未知溶液的荧光强度 I_f。

数据处理

（1）以各标准溶液的 I_f 为纵坐标、分别以邻羟基苯甲酸或间羟基苯甲酸的浓度为横坐标绘制标准曲线。得到回归方程。

（2）将 pH＝5.5 时未知溶液的荧光强度，代入邻羟基苯甲酸的线性回归方程，计算邻羟基苯甲酸在未知溶液中的浓度。

（3）将 pH＝12 时未知溶液的荧光强度与 pH＝5.5 时未知溶液的荧光强度的差值，代入间羟基苯甲酸的回归方程，计算未知溶液中间羟基苯甲酸的浓度。

思考与讨论

（1）λ_{ex}^{max} 和 λ_{em}^{max} 各代表什么？

（2）影响物质荧光强度的因素有哪些？

实验 10　荧光光谱法测定茶叶中铝离子含量

实验目的

（1）掌握直接荧光光谱法测定铝离子的基本原理和方法。

（2）进一步熟悉荧光光谱测定、溶剂萃取等基本操作。

实验原理

茶树是一种聚铝性植物，茶叶中铝含量对茶树生长及茶叶品质有重要影响。目前测定铝常用的方法有分光光度法和荧光分光光度法，其中荧光分光光度法灵敏度高，应用更为广泛。

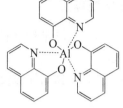

铝离子本身无荧光，但是它可与8-羟基喹啉反应形成可发射荧光的络合物8-羟基喹啉铝（8-Hydroxyquinoline aluminum，$C_{27}H_{18}AlN_3O_3$），其结构式如图3-5所示。该络合物为脂溶性物质，可被氯仿有效地从水相中萃取出来。萃取液以荧光法进行测定，最大激发波长和最大发射波长为390nm和510nm。因为络合物的荧光发射强度与其浓度成正比，依此可建立测定铝离子的荧光光谱法。

图3-5　8-羟基喹啉铝的结构式

仪器与试剂

（1）**仪器**。LS 55荧光光谱仪，石英池，分液漏斗（125mL），移液管，容量瓶，消解罐，烘箱。

（2）**试剂**

① 2.0μg/mL铝离子储备液：溶解1.760g硫酸铝钾［$Al_2(SO_4)_3 \cdot K_2SO_4 \cdot 24H_2O$］于20mL水中，滴加1：1硫酸至溶液澄清，移至100mL容量瓶中，用蒸馏水稀释至刻度并摇匀。准确移取所得溶液2.00mL至1000mL容量瓶中，用蒸馏水稀释至刻度并摇匀制得2.0μg/mL铝离子储备液。②2% 8-羟基喹啉溶液：溶解2g的8-羟基喹啉于6mL冰醋酸中，用水稀释至100mL制得8-羟基喹啉溶液。③每升含醋酸铵200g及浓氨水70mL的缓冲溶液。④硝酸，高氯酸，6mol/L氨水，氯仿，茶样。

实验步骤

（1）**系列标准溶液的配制**。取6只50mL容量瓶，分别加入0mL、10.0mL、20.0mL、30.0mL、40.0mL和50.0mL铝离子储备液，用水稀释至刻度，摇匀。

（2）**荧光络合物的生成与萃取**。取6个125mL分液漏斗，先各加入水45mL，再分别加入以上标准溶液各5.00mL。沿壁往每个漏斗加入8-羟基喹啉溶液和缓冲溶液各2mL。摇匀反应5min后，以氯仿萃取2次，每次10mL。有机相通过干燥脱脂棉滤入50mL容量瓶中，并以少量氯仿洗涤脱脂棉，洗液并入容量瓶中，以氯仿稀释至刻度并摇匀。

（3）**激发光谱和发射光谱的绘制**。设定激发波长为390nm，在450～600nm间扫描发射光谱，设定发射波长为510nm，在330～460nm的波长范围内扫描激发光谱。在激发光谱和发射光谱上分别找出最大激发波长和最大发射波长。

（4）**标准溶液荧光的测量**。以最大激发波长的光激发试样，对各标准溶液在450～600nm的波长范围内扫描荧光光谱，记录其在最大发射波长处的荧光强度。每种溶液重复扫描三次，取其平均值。

（5）**未知试样的处理和测定**

① 精确称取0.5000g茶样（每样两份），放入消解罐中，加入HNO_3-$HClO_4$（4：1）于400℃加热消解至溶液透明，将透明液转到50mL容量瓶，用去离子水反复多次冲洗消解

管，稀释至刻度。从中吸取 5mL 溶液至 25mL 容量瓶中，用去离子水稀释至刻度。

② 再从中吸取 5mL 茶样溶液，放入 125mL 分液漏斗，用 6mol/L 的氨水调至中性后，按第（2）步继续处理样品。

③ 依照第（4）步条件测定其荧光强度。

数据处理

（1）以标准溶液的浓度为横坐标，荧光强度为纵坐标，得到标准曲线及线性回归方程。

（2）根据茶样的荧光强度，以回归方程计算其铝离子含量。

思考题

（1）氯仿萃取液为何要以干燥脱脂棉过滤？

（2）分液漏斗旋塞处是否可用凡士林处理？为什么？

实验 11　阿司匹林片剂中乙酰水杨酸的荧光测定

实验目的

（1）熟悉荧光光度计的使用。

（2）掌握固体样品的处理方法。

实验原理

阿司匹林片剂是一种常见的止痛药。其主要成分是乙酰水杨酸，此外还含有其他成分，如黏合剂和缓冲剂。本实验中，将阿司匹林片剂溶于水，并通过加入氢氧化钠将其转化为水杨酸根离子（图 3-6）。在约 310nm 激发后，水杨酸根离子在约 400nm 处发出强烈荧光。制备一系列水杨酸根离子的标准溶液，然后测量标准溶液和样品溶液的荧光强度。采用标准曲线法测定样品溶液中水杨酸根离子的浓度，即可计算阿司匹林片剂中乙酰水杨酸的百分含量。

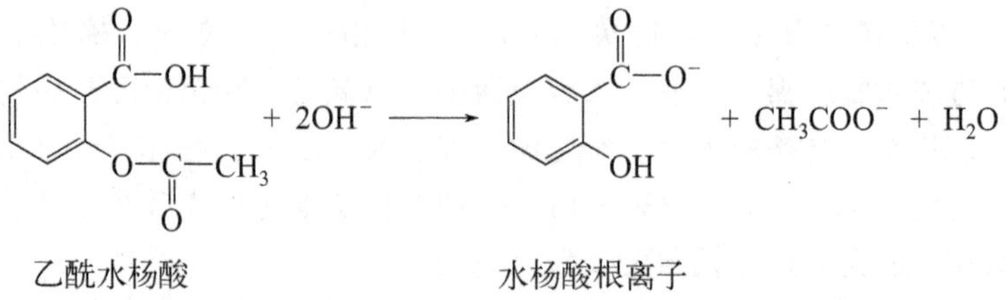

图 3-6　碱性条件下乙酰水杨酸转化为水杨酸根离子

仪器与试剂

（1）**仪器**。LS 55 荧光光谱仪，石英池（1cm）。

（2）**试剂**。阿司匹林片，100μg/mL 水杨酸原液，1mol/L 氢氧化钠溶液。

实验步骤

(1) 样品的配制

① 将阿司匹林片剂放在干净的研钵中，磨成粉末状。称取 0.005g 粉末（精确至 0.001mg）于 100mL 烧杯中，并用热水（80℃）将其溶解。将滤纸折叠后放入玻璃漏斗中，并将漏斗插入 100mL 容量瓶的瓶口，使滤液流入容量瓶中。用去离子水将溶液稀释至刻度并摇匀。

② 准确移取所得溶液 1.00mL 至 100mL 容量瓶中，用去离子水稀释至刻度并摇匀。

③ 在 3 个 50mL 容量瓶中，移取 2.00mL 氢氧化钠溶液至每一个容量瓶中，分别加入 5.00mL 的阿司匹林片剂稀释溶液。用去离子水将 3 个容量瓶稀释至刻度并摇匀。

(2) 标准溶液的配制

① 用所提供的 100μg/mL 水杨酸原液准确配制 1μg/mL 的水杨酸溶液。

② 取 5 只 50mL 容量瓶，先移取 2.00mL 氢氧化钠溶液至每一个容量瓶中，然后依次分别加入 1.00mL、2.00mL、3.00mL、4.00mL、5.00mL 的 1μg/mL 水杨酸溶液，用去离子水定容。

(3) 扫描荧光光谱。设置激发波长为 310nm，取 1 号标准溶液，在 320～600nm 间扫描发射光谱。用得到的最大发射波长，在 200～380nm 间扫描激发光谱，确定水杨酸的最大激发波长和最大发射波长。

(4) 测定未知样品

① 在最佳激发和发射波长下，依次测定所有标准溶液的荧光强度。

② 同样条件下，测量样品溶液的荧光强度。

数据处理

（1）以水杨酸标准溶液的荧光强度为纵坐标，其浓度为横坐标，绘制标准曲线，进行线性回归，拟合回归方程。

（2）根据未知试样的荧光强度，以回归方程计算稀释后样品溶液中水杨酸根离子的浓度。

（3）根据稀释倍数计算所称量的片剂粉末中乙酰水杨酸的质量（mg）。

（4）根据所求得的片剂粉末中乙酰水杨酸的质量计算片剂中乙酰水杨酸的百分含量。报告样品的平均值和标准偏差。

思考题

（1）画出荧光分析仪的部件组成方框图。

（2）查阅资料，讨论乙酰水杨酸的其他测定方法。

实验 12　苯甲酸、乙酸乙酯的红外光谱测定

实验目的

（1）学习红外光谱法的基本原理及仪器构造。

（2）了解红外光谱法的应用范围。

（3）初步掌握固态和液态样品的制备方法。

实验原理

红外光谱反映分子的振动情况。当用一定频率的红外光照射某物质时，若该物质的分子中某基团的振动频率与之相同，则该物质就能吸收此种红外光，使分子由振动基态跃迁到激发态。当用不同频率的红外光通过待测物质时，就会出现不同强弱的吸收现象。

由于不同化合物具有其特征的红外光谱，因此可以用红外光谱对物质进行结构分析。因为是光的吸收过程，根据比尔定律也可以对物质进行定量分析。

仪器与试剂

（1）仪器。FTIR Nicolet iS50 傅里叶变换红外光谱仪，油压式压片机，红外干燥灯。

（2）试剂。无水乙醇、乙酸乙酯、苯甲酸均为分析纯，光谱纯 KBr，玛瑙研钵，盐片，某未知物。

实验步骤

（1）固体样品苯甲酸的红外光谱测定。取约 1mg 苯甲酸样品于干净的玛瑙研钵中，加约 100mg KBr 粉末，在红外灯下研磨成粒度约 $2\mu m$ 细粉后，移入压片模中，将模子放在油压机上，在 16MPa 的压力下维持 2min，放气去压，取出模子进行脱模，可获得一片直径为 13mm 的半透明盐片。将盐片装在样品池架上，即可进行红外光谱测定。

（2）液体样品乙酸乙酯的红外光谱测定。在一块干净抛光的 NaCl 或 KBr 片上滴加一滴乙酸乙酯样品，压上另一块盐片，将其置于样品池架上，即可进行红外光谱测定。

（3）未知物的红外光谱测定。根据提供的未知物，确定样品制备方法并测定其红外光谱。

数据处理

（1）对苯甲酸及乙酸乙酯的特征谱带进行归属。

（2）推测未知物可能的结构。

注意事项

（1）在红外灯下研磨固体样品，防止吸潮。

（2）盐片应保持干燥透明，每次测定前均应用无水乙醇及滑石粉抛光（红外灯下），切勿水洗。

思考题

（1）红外吸收光谱仪和紫外可见吸收光谱仪有哪些相同点和不同点？

（2）测试红外光谱时，分散剂一般常用 NaCl 和 KBr，它们适用的波数范围各为多少？

实验 13　火焰原子吸收光谱法测定自来水的钙镁硬度
（见视频 3）

视频 3

实验目的

（1）熟悉原子吸收光谱仪的结构及其使用方法。

（2）掌握原子吸收光谱法进行元素定量分析的基本方法（标准曲线法、标准加入法）。

（3）掌握原子吸收光谱法元素灵敏度的计算方法。

实验原理

水的硬度定义为水中碱土金属离子的总浓度。由于 Ca^{2+} 和 Mg^{2+} 的浓度通常要比其他碱土金属离子浓度高，水的总硬度可以等同于水中钙、镁离子的总含量。分硬度是指每种碱土金属离子的个体浓度。常以每升水中所含有的 $CaCO_3$ 的含量（mg）表示总硬度。硬度超过每升 60mg 的水被认为是硬水。水的硬度可通过许多方法测定，如 EDTA 滴定法、原子吸收光谱法（AAS）和原子发射光谱法（AES）等。

AAS 常用的原子化系统有三种，即火焰原子化系统、石墨炉原子化系统和低温原子化系统。其中火焰原子化操作简单、分析速度快、分析成本低，对大多数元素有较高的灵敏度，应用较为广泛。本实验采用火焰原子吸收分光光度法，测定自来水中 Ca^{2+} 和 Mg^{2+} 的浓度。

火焰原子吸收光谱法是基于待测元素的基态原子蒸气（来源于火焰原子化器）对从光源中辐射出的待测元素的特征谱线的吸收程度而建立的一种定量分析方法。在一定条件下，该吸收程度与试液待测元素的浓度成正比，即 $A = Kc$。采用标准曲线法或标准加入法，可以算出元素的含量。

采用标准曲线法时，先配制一系列待测元素的标准溶液，分别测出它们的吸光度 A，以 A 对 c 作图，经线性回归得到标准曲线和线性方程。在同样测量条件下，测出待测试液的吸光度 A_x，从线性回归方程即可计算待测元素的浓度 c_x。

采用标准加入法时，一般是量取 1、2、3、4、5 共 5 份等量的待测试液于一定体积的容量瓶中，在其中 2～4 份中分别加入一定浓度梯度的待测元素的标准溶液，定容后分别测定其吸光度。绘制吸光度对待测元素加入量的曲线，将此曲线外推，与浓度坐标的交点即为稀释后的试样中待测元素的含量。

在火焰 AAS 中，测定元素的灵敏度定义为产生 1% 吸收（$A = 0.0044$）时被测物的质量浓度。根据定义，元素的灵敏度 S 为：

$$S = \frac{c \times 0.0044}{A}$$

式中，c 为试液的质量浓度；A 为试液的吸光度。显然，S 越小，元素测定的灵敏度越高。

仪器与试剂

（1）**仪器**。GGX-810 型原子吸收光谱仪，钙空心阴极灯，镁空心阴极灯，空气压缩机，乙炔钢瓶。

（2）**试剂**。100mg/L 钙标准储备溶液，50mg/L 镁标准储备溶液，自来水。

实验步骤

（1）**仪器参数的设置**。按最佳仪器条件设置仪器参数。

（2）**标准溶液的配制**

① 钙标准溶液系列：配制浓度为 2.00mg/L、4.00mg/L、6.00mg/L、8.00mg/L、10.00mg/L 的钙标准溶液于 5 个 50mL 容量瓶中。

② 镁标准溶液系列：配制浓度为 0.25mg/L、0.50mg/L、0.75mg/L、1.00mg/L、1.25mg/L 的镁标准溶液于 5 个 50mL 容量瓶中。

（3）**硬度测定**

① 在选定的仪器工作条件下，以超纯水为空白，分别测定钙、镁系列标准溶液的吸光度。

② 取自来水样，以超纯水为空白，分别测定其中钙、镁的吸光度值。

（4）**关机**。实验结束后用超纯水喷洗原子化系统 2min，按操作程序关机。

数据处理

（1）将所得钙、镁离子的实验数据制成表格，并以金属元素的浓度为横坐标，所测得的吸光度值为纵坐标，绘制标准曲线。

（2）计算自来水中钙、镁离子的浓度及水的硬度，并判断水的软硬程度。

（3）计算钙、镁元素的检测灵敏度。

思考题

（1）简述原子吸收分光光度法的基本原理。

（2）原子吸收分光光度分析为何要用待测元素的空心阴极灯做光源？

实验 14   火焰原子吸收光谱法测定黄铜中的铜和铅

实验目的

（1）进一步熟悉原子吸收分光光度计的操作。

（2）了解黄铜预处理过程。

实验原理

黄铜是一种主要由铜和锌组成的合金。铜和锌的比例不同，黄铜种类也不同。铜的含量在 55% 到 95%（按质量计）不等，大多数黄铜是由约 67% 的铜和 33% 的锌组成。

铅通常以约 2% 的浓度添加到黄铜中，铅的添加提高了黄铜的切削性能。在该实验中，使用标准曲线方法通过火焰原子吸收光谱法（AAS）测定黄铜中的铜和铅的含量。

仪器与试剂

（1）**仪器**。GGX-810 型原子吸收光谱仪，Pb 和 Cu 空心阴极灯（分别测定铅和铜在

283.3nm 和 324.8nm 处的吸光度），空气压缩机，乙炔钢瓶。

（2）**试剂**。浓 HNO_3，铅弹，铜箔，黄铜。

实验步骤

（1）铅储备溶液的配制。准确称量 0.5g 左右的铅弹于 150mL 烧杯中（不要在烘箱中干燥），在通风橱中加入 20mL 去离子水和 20mL 浓 HNO_3，盖上表面皿并微热煮沸，直到金属溶解，溶液呈无色。如果形成白色沉淀，则冷却溶液并加入 20mL 去离子水。用少量去离子水冲洗表面皿和烧杯内壁，再煮沸 10min。溶液冷却后定量转移至 1000mL 容量瓶中，稀释至刻度并摇匀。Pb 的最终浓度约为 500mg/L。

（2）铜储备溶液的配制。准确称量约 0.5g 铜箔样品，在通风橱中加入 50mL 稀释的 HNO_3（3mol/L），盖上表面皿，在电热板上加热，待 Cu 全部溶解后，继续微沸 10min（避免溶液蒸干）。将样品冷却至室温并定量转移至 1000mL 容量瓶中，用去离子水稀释至刻度，摇匀。

（3）黄铜样品储备溶液的配制。准确称取约 1.0g 黄铜样品于 150mL 烧杯中，在通风橱中加入 10mL 水，然后加入 15mL 浓 HNO_3，待气体放出减慢后，再加入 10mL 水。盖上表面皿，在电热板上加热溶解，待所有黄铜溶解后继续保持溶液微沸 20min。冷却后，蓝色溶液中可能存在蓬松的白色沉淀物，此沉淀物是水合氧化锡（H_2SnO_3）。小心地将溶液通过锥形滤纸过滤到干净的烧杯中，然后将溶液定量转移至 500mL 容量瓶中，加去离子水稀释至刻度，摇匀。

（4）样品分析

① 使用 Pb 和 Cu 的储备溶液配制 Pb 和 Cu 的标准溶液（在 0 至 50mg/L 之间）。

② 分别测量 Pb 和 Cu 系列标准溶液的吸光度。

③ 测量样品中 Cu 和 Pb 的吸光度，注意调整样品中黄铜的浓度，使 Cu 和 Pb 的浓度落在标准曲线线性范围内。

数据处理

（1）绘制平均吸光度与金属浓度的标准曲线，包括每个点的误差线（标准偏差）。

（2）使用标准曲线确定未知样品中 Pb 和 Cu 的浓度。并以 95％置信度，报告样品中 Pb 和 Cu 含量的置信区间（准确到 0.01％）。

注意事项

（1）记录 Cu、Pb 和样品的质量，精确到 0.1mg。

（2）取上述样品称量时戴上手套，避免指纹等的污染，并且在称重前不要在烤箱中烤干。

思考题

（1）如果步骤（3）中有白色沉淀产生，则怎样采用重量分析法定量测定样品中的 Sn？

（2）查找一篇用其他方法测定黄铜中 Cu 含量的文献，并说明文献中定量测定的方法。

实验 15　石墨炉原子吸收光谱法测定土壤中镉的含量

实验目的

（1）学会使用石墨炉原子吸收光谱仪。

（2）了解土壤中镉测定的前处理技术。

（3）掌握镉测定的两种定量方法（标准曲线法和标准加入法）。

实验原理

原子吸收光谱法常用的原子化系统有：火焰原子化系统、石墨炉原子化系统和低温原子化系统。不同类型原子化系统直接影响元素分析的灵敏度、检出限、精密度和线性范围。火焰原子化操作简单、分析速度快、分析成本低，对大多数元素有较高的灵敏度，应用较为广泛。但火焰原子化所需试样溶液体积大，原子化效率低。

石墨炉原子化是采用电流加热石墨炉原子化器，使之达到 2000℃ 以上的高温，从而使样品达到原子化的技术。石墨炉升温程序按干燥、灰化、原子化、除残 4 步完成。各阶段温度、温度保持时间、升温方式按试样组成及分析元素设置。干燥是为了脱溶剂避免在灰化、原子化时试样飞溅。灰化的作用是除去易挥发的基体和有机物，以减少分子吸收。原子化阶段温度最高，分析元素蒸发并解离为基态原子蒸气，这时惰性保护气停气，以延长基态原子在石墨管中的停留时间，从而提高方法的分析灵敏度。除残是在高温时除去留在石墨管中的基体残留物，消除记忆效应，为下一次测定做准备。

石墨炉原子吸收光谱法（GFAAS）精密度较差，操作也较为复杂。但该方法的优点是：试样用量少，原子化效率几乎达到 100％，分析绝对灵敏度高，是土壤中微量镉的理想分析方法。

仪器与试剂

（1）**仪器**。原子吸收光谱仪（AA800，美国 PE 公司），镉空心阴极灯，无油空气压缩机。

（2）**试剂**。高纯氩气，盐酸、硝酸、氢氟酸、高氯酸均为优级纯，镉粉（99.99％），超纯水，土壤样品。

实验步骤

（1）**土壤试液的制备**。称取 0.50～1.0g 土样于 25mL 聚四氟乙烯坩埚中，用少许水润湿，加入 10mL HCl，在电热板上加热（<450℃）消解 2h，然后加入 15mL HNO_3 继续加热至溶解物剩余约 5mL 时，再加入 5mL HF 并加热分解去除硅的化合物，最后加入 5mL $HClO_4$ 加热至消解物呈淡黄色时，打开盖，蒸至近干。取下冷却，加入 1∶5 的 HNO_3 1mL 微热溶解残渣，移入 50mL 容量瓶中，定容。同时进行全程序试剂空白实验。

（2）**镉标准溶液的配制**。镉标准储备液：称取 0.5000g 金属镉粉，微热溶于 25mL 1∶5 的 HNO_3 中。冷却，移入 500mL 容量瓶，用去离子水稀释并定容。此溶液每毫升含镉 1.0mg。

镉标准使用液：吸取 10.0mL 标准储备液于 100mL 容量瓶中，用水稀释至标线，摇匀备用。吸取 5.0mL 稀释后的标液于另一 100mL 容量瓶中，用水稀释至标线即得每毫升含 5μg 镉的标准使用液。

（3）**仪器预热**。依次打开氩气开关、循环冷却水器（18～22℃）、石墨炉电源、原子吸收光谱仪电源、仪器工作站，启动软件，开启元素灯，预热 40min。

（4）**测定条件设置**。镉空心阴极灯的测定波长是 228.8nm，通带宽度 1.0nm，灯电流 1mA，氩气流速 0.2L/min，进样体积 20μL，石墨炉升温程序（供参考）如表 3-3 所示。

表 3-3　土壤中镉测定的石墨炉升温程序

步骤	温度/℃	时间/s	升温梯度/(℃/min)	气体流量/(L/min)
干燥	100	30	10	0.2
灰化	300	20	150	0.2
原子化	900	30	0	0
净化	2500	30	0	0.2

（5）**样品测定**

① 标准曲线法。分别吸取镉标准使用液 0mL、0.50mL、1.00mL、2.00mL、3.00mL、4.00mL 于 6 个 50mL 容量瓶中，用 0.2％HNO₃ 溶液定容、摇匀。此标准系列分别含镉 0μg/mL、0.05μg/mL、0.10μg/mL、0.20μg/mL、0.30μg/mL、0.40μg/mL，依次测其吸光度。同样条件下测定土壤样品中镉的吸光度。

② 标准加入法。分别移取试样溶液 5.0mL 于 5 个 10mL 容量瓶中，依次分别加入镉标准使用液（5.0μg/mL）0mL、0.50mL、1.00mL、1.50mL、2.00mL，用 0.2％HNO₃ 溶液定容，分别测定各溶液吸光度。

数据处理

（1）以标准曲线法中各物质吸光度作为纵坐标，各物质浓度为横坐标，得到标准曲线和线性回归方程。将土壤样品吸光度（扣除空白吸光度）代入回归方程，计算得到土壤中镉含量。

（2）以加入的标准溶液浓度为横坐标，测得的吸光度为纵坐标，绘制 A-c 标准曲线。外延曲线与横坐标相交，原点与交点的距离，即为稀释后待测镉离子的浓度。最后计算土壤中镉含量。

注意事项

（1）土样消化过程中，最后除 $HClO_4$ 时必须防止将溶液蒸干，不慎蒸干时 Fe、Al 盐可能形成难溶的氧化物而包藏镉，使结果偏低。

（2）镉的测定波长为 228.8nm，该分析线处于紫外光区，易受光散射和分子吸收的干扰。特别是在 220.0～270.0nm，NaCl 有强烈的分子吸收，覆盖了 228.8nm 线。另外，Ca、Mg 的分子吸收和光散射也十分强。这些因素皆可造成镉的表观吸光度增大，为消除基体干扰，可在测量体系中加入适量基体改进剂，如在标准系列溶液和试样中分别加入 0.5g $La(NO)_3 \cdot 6H_2O$ 此法适用于测定受镉污染土壤中的镉含量。

（3）高氯酸的纯度对空白值的影响很大，直接关系到测定结果的准确度，因此必须注意全过程空白值的扣除，并尽量减少加入量以降低空白值。

思考题

（1）不同的石墨炉程序升温对测定结果有何影响？

（2）土壤未消解完全对测定结果有何影响？

实验 16　石墨炉原子吸收光谱法测定硒补充剂中的硒

实验目的

（1）进一步掌握石墨炉原子吸收光谱仪的操作技术。

（2）掌握基体改进剂的作用。

实验原理

硒（Se）是人体必需的微量元素。它能保护人体组织免受氧化应激，加强对感染的防御，以及调节生长发育。慢性硒缺乏可能使我们更易受到病毒感染、癌症、心血管疾病、甲状腺功能障碍等的侵袭。缺硒人群可采用硒补剂按人体需要进行定量、精确补硒。

硒是一种挥发性元素。本实验采用低灰化温度、加入基体改进剂和石墨炉程序升温，提高分析的灵敏度和稳定性。

仪器与试剂

（1）仪器。 TAS-990AFG 石墨炉原子吸收光谱仪，硒空心阴极灯，容量瓶。

（2）试剂。 浓 HNO_3，1000mg/L Se 储备液，$Ni(NO_3)_2 \cdot 6H_2O$，硒膳食补充片剂。

实验步骤

（1）溶液的制备

① 称取适量的 $Ni(NO_3)_2 \cdot 6H_2O$ 溶于去离子水中，并稀释至 50mL，得到 5%Ni 溶液。

② 移取适量 Se 原液（1000mg/L），用去离子水稀释至 500mL，得到 1mg/L Se 储备液。

③ 移取适量 1mg/L 的 Se 储备溶液，加入 0.5mL 浓 HNO_3 和 2mL 5%Ni 溶液，准确配制浓度分别为 0μg/L、5μg/L、10μg/L、20μg/L 和 40μg/L 的 100mL 标准溶液。

（2）样品制备。 称取一粒硒膳食补充片剂放入 250mL 玻璃烧杯中，加入 50mL 去离子水，加入足够的浓 HNO_3，使酸浓度为 1%。然后在 95℃下加热 1h，待冷却至室温后转移到容量瓶中，用去离子水稀释至 100mL。

（3）仪器工作条件。 灯电流 9.0mA，光谱带宽 1.0nm，分析线 196.0nm，样品溶液进样量 10μL。石墨炉升温程序见表 3-4。

表 3-4　测定硒的石墨炉升温程序

步骤	温度/℃	时间/s	升温梯度/(℃/min)	气体流量/(L/min)
干燥 1	90	10	5	3.0
干燥 2	105	20	3	3.0
干燥 3	300	10	50	3.0

步骤	温度/℃	时间/s	升温梯度/(℃/min)	气体流量/(L/min)
灰化	1100	5	50	3.0
原子化	2100	4	1400	0
除残	2300	4	500	3.0

（4）未知样品分析

① 打开 TAS-990AFG 石墨炉原子吸收光谱仪和计算机，设置分析方法。

② 依次测定上述制备的含硒标准溶液的吸光度。

③ 根据标准曲线的线性范围，对未知样品进行适当稀释。测定样品和空白溶液的吸光度，重复测定三次。

数据处理

（1）绘制标准曲线，并计算未知样品中 Se 的浓度。

（2）计算硒膳食补充剂中 Se 的浓度，并将结果与制造商的结果进行比较。

思考题

（1）本实验中硝酸镍溶液的作用是什么？

（2）设计采用标准加入法测定实际样品中硒含量的实验方案。

实验 17 电感耦合等离子原子发射光谱分析法测定废水中的镉、铬含量

实验目的

（1）掌握 ICP-AES 光谱仪的构成、基本操作技术。

（2）掌握 ICP 光源的工作原理。

（3）掌握多元素混合标准溶液（多标溶液）的配制方法。

实验原理

原子发射光谱法是根据元素的气态原子或离子受激发后所发射的特征波长及其强度来测定物质组成和含量的分析方法。在激发光源中，被测定物质被蒸发、解离、电离、激发，产生辐射。目前常用的光源有电弧（arc）、电火花（spark）和电感耦合等离子（inductively coupled plasma，ICP）等。其中 ICP 光源具有激发能力强、稳定性好、基体效应小、检出限低等优点。同时，ICP 光源无自吸现象，标准曲线的线性范围宽。

电感耦合等离子原子发射光谱分析法（ICP-AES 或 ICP-OES）是一种能同时进行多元素检测的发射光谱技术。它将试样在等离子体光源中激发、使待测元素发射出特征波长的辐射，经过分光，根据其波长所在位置进行定性分析，测量其谱线发射强度进行定量分析。由于元素谱线的发射强度和被测元素浓度成正比，因此可以用标准曲线法、标准加入法及内标法进行试样中待测元素的定量分析。

废水中的镉和铬是水中的第一类污染物，对人类危害极大。本实验在对水样预处理后，采用标准曲线法同时测定废水中 Cd 和 Cr 的含量。

仪器与试剂

（1）仪器。安捷伦 5100 ICP-OES 原子发射光谱仪，烧杯，容量瓶，电炉，超声仪。

（2）试剂。100mg/L 镉标准储备液，100mg/L 铬标准储备液，硝酸（HNO_3），高氯酸（$HClO_4$），工业废水。

实验步骤

（1）**工作参数设置**。分析线波长：Cd 226.502nm，Cr 267.716nm；入射功率：1kW。氩冷却气流量：12~14L/min，氩辅助气流量：0.5~0.8L/min，氩载气流量：1.0L/min。

（2）**标准溶液的配制**。配制浓度为 0.1mg/L、0.5mg/L、1.0mg/L、5.0mg/L、10mg/L 镉和铬的系列双元素标准溶液。

（3）**水样的预处理**

取 50mL 水样于 100mL 烧杯中，在电炉上微热至 10mL 左右，加入混合酸（HNO_3：$HClO_4$=4:1）10mL，蒸至近干，冷却后加入 5mL HNO_3（1:1）和少量去离子水至约 25mL，置于超声波中超声 0.5h，溶解残渣，过滤后滤液转入 50mL 容量瓶中，定容，摇匀。

（4）**样品测定**

① 按照基本操作步骤完成仪器开机及点燃 ICP 炬。

② 在元素分析表中选择分析元素、分析线波长及最佳工作条件。

③ 依次喷入不同浓度的混合标准溶液，测定其发射强度，仪器自动绘制标准曲线。

④ 喷入工业废水试液，采集测试数据。仪器自动计算测试结果。

⑤ 按照关机程序，退出分析程序、进入主菜单、关蠕动泵、关气路、关 ICP 电源及计算机系统、最后关冷却水。

注意事项

（1）测试完毕后，进样系统用去离子水喷洗 3min 后关机，以免试样沉积在雾化器口和石英炬管口。

（2）等离子体发射很强的紫外光，易伤眼睛，应通过有色玻璃防护窗观察 ICP 炬。

思考题

（1）为什么 ICP 光源能够提高原子发射光谱分析的灵敏度和准确度？

（2）简述 ICP-AES 与火焰原子吸收光谱法的异同点。

实验 18 ICP-AES 法测定纯锌试样中的杂质元素

实验目的

（1）进一步掌握 ICP-AES 的基本工作原理。

（2）进一步掌握采用 ICP-AES 同时测定多种元素的分析方法。

实验原理

纯锌试样通常含有 Pb、Cd、Fe、Cu、Sn、Al、As、Sb 等多种杂质元素，其杂质含量是确定产品质量的一个关键指标。采用分光光度法或火焰原子吸光法测定这些元素的含量，既麻烦又耗时。电感耦合等离子原子发射光谱分析法（ICP-AES）可同时进行多元素分析，具有分析速度快、灵敏度高、稳定性好、线性范围宽、基体干扰小等优点。采用 ICP-AES 分析方法测定纯锌试样中的杂质元素，不仅可以大大提高分析效率，还可以使分析结果更加准确可行。

仪器与试剂

（1）**仪器。**安捷伦 5100 ICP-OES 原子发射光谱仪，烧杯，电热板，容量瓶。

（2）**试剂。**1.000mg/mL 多元素标准储备液，浓盐酸（AR），超纯水，纯锌试样。

实验步骤

（1）**纯锌试样溶液的制备。**用电子天平准确称取 0.5g 左右的纯锌试样于 100mL 烧杯中，加入 10mL 1:1 盐酸，盖上表面皿，在电热板上加热。待锌粒溶完后，将溶液蒸发近干，用洗瓶冲洗表面皿和烧杯内壁。冷却后将其转移到 25mL 容量瓶中，用 5% 的 HCl 超纯水溶液定容，摇匀。

（2）**多元素标准溶液的制备。**配制浓度为 0.1μg/mL、0.5μg/mL、1.0μg/mL、5.0μg/mL、10.0μg/mL 多元素系列标准溶液，所有溶液都要用 5% 的 HCl 超纯水溶液进行定容。

（3）**仪器参数设置。**高频功率：1150W；冷却气流量：12L/min；辅助气流量：0.5L/min；载气流量：1.0L/min；蠕动泵转速：50r/min。

（4）**样品分析**

① 按照 5100 ICP-OES 原子发射光谱仪的基本操作步骤完成准备工作，开机及点燃等离子体。

② 测定多元素标准溶液的发射强度，仪器自动绘制各标准曲线。

③ 喷入制备好的纯锌试样溶液，仪器自动给出测定结果。根据测定结果，计算杂质元素的质量分数（%）。

$$w_x = \frac{\rho V \times 10^{-6}}{m} \times 100\%$$

式中，ρ 为测定纯锌试样中杂质元素的质量浓度，单位为 μg/mL；V 为溶液的体积，单位为 mL；m 为试样的质量，单位为 g。

④ 确认所有分析工作完成后，用 5% 的 HCl 超纯水溶液冲洗气路 5min，再用超纯水冲洗 5min，然后熄灭等离子体。5min 后关冷却水，待 CID 检测器温度升至室温后关闭氩气。最后关闭排风。

注意事项

（1）溶样过程中要等溶液冷却后再转移到容量瓶中定容，以免定容体积产生误差。

（2）如果试样盐分较高应随时观察雾化情况，防止雾化器口堵塞。试样未完全溶解严禁上机测定。

思考题

(1) 简述 ICP 产生的原理及过程。

(2) 如何选择多元素同时分析仪器的工作参数？

实验 19 微波消解-ICP-AES 法同时测定火腿中多种微量元素

实验目的

(1) 掌握用微波消解技术进行样品前处理的方法。

(2) 了解微波消解仪的使用。

(3) 掌握 ICP-AES 同时检测多元素的定性和定量分析方法。

实验原理

金华火腿是浙江省著名特产。火腿肉色红润、香气浓郁，以"色、香、味、形"四绝驰名中外，其中尤以产地金华的火腿为盛。火腿营养丰富，除含有大量的蛋白质、脂肪外，还含有多种人体必需的元素如钾、钙、铁等，研究火腿中微量元素含量，对于分析评价火腿的质量和营养价值具有重要的作用。

微波消解是一种样品前处理技术，该技术在高压的密闭容器内消解样品，该方法具有待测元素不受污染，易挥发元素不易损失，样品消解完全、快速，节省试剂等优点。微波消解前处理联合 ICP-AES 技术，可以实现对环境学、高纯材料、地质学、核科学、生物学、医药学、化学计量学、农业和食品学研究中的痕量和超痕量元素进行高灵敏多元素快速测定。

本实验首先采用微波消解技术对火腿样品进行前处理后，采用电感耦合等离子体原子发射光谱法（ICP-AES）的标准曲线法测定火腿中钾（K）、钙（Ca）、钠（Na）、镁（Mg）各元素的含量。

仪器与试剂

(1) 仪器。安捷伦 5100 ICP-OES 原子发射光谱仪，ETHOS 微波消解仪。

(2) 试剂。干腌火腿，优级纯硝酸，1000mg/L K、Ca、Na、Mg 标准储备液。

实验步骤

(1) 样品处理

① 将火腿肉用绞肉机绞碎，备用。

② 准确称取样品 0.2000g 左右于聚四氟乙烯消解罐中，加入一定量硝酸，将消解罐拧紧，设置合适的参数进行微波消解。待程序结束后自动冷却，取出消解罐，用电热板加热赶酸，冷却后，移入 25mL 比色管中，用去离子水定容。同时做 3 个空白试样。微波消解仪工作条件见表 3-5。

表 3-5　微波消解仪的工作条件

步骤	温度变化/℃	时间/s	功率/W
1	室温~120	300	500
2	保持 120	300	500
3	120~195	600	800
4	保持 195	600	800
5	冷却	600	0

（2）**仪器工作条件**。按照 5100ICP-OES 原子发射光谱仪的基本操作步骤完成准备工作，开机及点燃等离子体。部分仪器工作参数如下：功率为 1.15kW，载气流量为 0.75L/min，辅助气流量为 1.5L/min，冷却气流量为 15L/min，观察位置自动优化，观测方向为轴向。各元素的分析谱线可按照表 3-6 选择测定波长。

表 3-6　各元素的分析谱线

测定元素	波长/nm	测定元素	波长/nm
K	766.491	Na	589.592
Ca	422.673	Mg	279.533

（3）**绘制标准曲线**。将标准储备液用 5% 的硝酸按 0.00μg/mL、5.00μg/mL、10.00μg/mL、20.00μg/mL、50.00μg/mL、100μg/mL 梯度进行配制。在选定的仪器条件下分别将不同浓度的多标溶液依次喷入仪器，采集数据，仪器自动得到各元素的标准曲线。

（4）**样品测定**。喷入配制好的火腿样品溶液，采集数据，仪器自动给出各元素的测定结果。

（5）**关机**。分析工作完成后，用 5% 的 HCl 高纯水溶液冲洗 5min，再用高纯水冲洗 5min，然后熄灭等离子体。5min 后关闭冷却水，待检测器温度升至室温后关闭氩气。最后关闭排风。

数据处理

根据测定结果计算每克火腿中各元素含量（μg/g）。

思考题

（1）采用 ICP-AES 进行多元素分析的优点是什么？

（2）样品前处理有哪些方法？

实验 20　电感耦合等离子体质谱法测定饮用纯净水中 24 种金属元素

实验目的

（1）了解 ICP-MS 工作的基本原理。

（2）了解 ICP-MS 的仪器结构和操作方法。

实验原理

电感耦合等离子体质谱法（inductively coupled plasma mass spectrometry，ICP-MS）是以电感耦合等离子作为离子源，以质谱进行检测的无机多元素分析技术。它是一种将 ICP 技术和质谱结合在一起的分析技术，能同时测定几十种痕量无机元素，可进行同位素分析、单元素和多元素分析，以及有机物中金属元素的形态分析。

ICP-MS 工作原理是样品由载气（氩气）带入雾化系统进行雾化后，以气溶胶形式进入等离子的轴向通道，在高温和惰性气体中被充分蒸发、原子化和离子化，产生的离子经过采样锥和截取锥进入真空系统，经过离子镜聚焦，由四极杆质谱仪依据质荷比进行分离。根据质谱峰的位置及元素浓度，进行试样中元素的定性和定量分析。

饮用纯净水是不含任何添加物、可直接饮用的水。饮用纯净水中金属元素超标直接影响人类的身体健康，监测饮用纯净水中金属元素的含量，保证饮用水的质量，对于维护消费者的身体健康有着重要意义。

常用的检测饮用水中金属元素的方法有原子吸收光谱法、原子荧光法、分光光度法、电感耦合等离子体发射光谱法和电感耦合等离子体质谱法等。其中，ICP-MS 由于其灵敏度高、检出限低、线性范围宽且能够实现多元素的同时测定等优点，是水质中多种金属元素同时测定的最佳选择。

仪器与试剂

（1）仪器。iCAP Q 电感耦合等离子体质谱仪，Milli-Q 超纯水系统。

（2）试剂。饮用纯净水来源于市售产品。硝酸（优级纯），Tl、Ba、Sb、Sn、Cd、Se、Cu、Ni、Co、Mn、V、Ti、Cr、Mo、B、Al、Zn、Fe、Mg、Ca、As、Sr、Hg、Pb 的多元素标准溶液（浓度为 100mg/L，国家标准物质中心）。

内标元素：Sc、Ge、Rh、Bi（浓度为 1000μg/L，国家标准物质中心）。

实验步骤

（1）样品前处理。量取纯净水 10mL，加入 1 滴硝酸，混匀，待测。

（2）ICP-MS 工作条件设置。射频功率为 1500W，等离子体气流量：15L/min；载气流量：0.80L/min；辅助气流量：0.40L/min；雾化室温度为 2℃，雾化器为同心雾化器，采样锥和截取锥分别为镍/铂锥，采集模式为跳峰。

（3）标准溶液的配制与内标的选择。取一定量的多元素标准储备液，用 2% 的硝酸逐级稀释，得到浓度范围为 2~50μg/L 的标准溶液。内标元素选择应该与待测元素质量数接近，分别选择 Sc、Ge、Rh、Bi 作为内标元素，用 2% 的硝酸溶液稀释至内标储备液浓度为 20μg/L。根据不同元素选择的内标物为：

① B、Mg、Al、Ca、Fe 选用 Sc 做内标。

② Ti、Zn、V、Cr、Mn、Co、Ni、Cu 选用 Ge 做内标。

③ As、Se、Sr、Mo 选用 Rh 做内标。

④ Cd、Sn、Sb、Ba、Hg、Tl、Pb 选用 Bi 做内标。

（4）样品测定。检查真空度、氩气压力以及循环冷却水等各参数正常后，点燃等离子体。待稳定后，优化仪器各参数。通过三通阀，将内标与待测液等体积进样。编辑并调入序列，进行样品分析。

数据处理

列出水样中痕量元素的含量。

注意事项

（1）所用器皿均用 $10\%HNO_3$ 浸泡 24h，然后用超纯水冲洗，备用。

（2）用该方法可检测砷、铜、铅、铁、锰、锌、镉、铝、汞、钡、硼、钼、钠、硒、银、镍、铊、铍、锑等 24 种元素。如果标样较少，也可以选择几种痕量元素进行检测测定。

思考与讨论

（1）常用的金属元素的检测测定方法有哪几种？比较它们的优缺点。

（2）采用电感耦合等离子体质谱法测定水中的痕量元素分析，有哪些质谱干扰？如何消除？

实验 21　直接电位法测定未知溶液的 pH
（见视频 4）

实验目的

（1）掌握用直接电位法测定溶液 pH 的原理。

（2）学会使用 pHS-3C 型 pH 计。

视频 4

实验原理

直接电位法根据测量组成电化学池的指示电极的电位值，从 Nernst 方程直接求得被测定物质活度（或浓度）的分析方法。水溶液的酸碱度的准确测定普遍采用直接电位法。IUPAC 推荐采用与标准缓冲溶液直接比较法进行测定。测定时，把玻璃电极与参比电极组成下列电池：

$$Hg，Hg_2Cl_2｜饱和 KCl ‖ 试液｜玻璃膜｜内参比溶液｜AgCl，Ag$$

该电池的电动势为

$$E_{电池}=E_{玻}-E_{SCE}+E_{液接}$$

式中，E_{SCE} 为参比电极的电位；$E_{玻}$ 表示玻璃电极的电位，它能反映待测物质的活度（浓度）信息；$E_{液接}$ 是通过盐桥的接触电位。一定条件下 E_{SCE} 和 $E_{液接}$ 为常数。而

$$E_{玻}=k-0.059pH$$

因此电池的电动势可简写为 $E_{电池}=K-0.059pH$（25℃）。

若上式 K 已知，则由测得的 E 值可求出被测溶液的 pH，但实际上 K 值不易求得，因此在实际测量中，用已知的标准缓冲溶液做基准，比较待测溶液和标准溶液两个电池的电动势来确定待测溶液 pH，该方法称直接比较法，或两点标定法。其中 0.059V/pH（或 59mV/pH）称 pH 玻璃电极响应斜率（25℃），理想的 pH 玻璃电极在 25℃时其斜率应为

59mV/pH，但实际上由于制作工艺等的差异，每个 pH 玻璃电极其斜率可能不同，须用实验方法来测定。

仪器与试剂

（1）**仪器**。pHS-3C 型酸度计，雷磁 E-201-C 型 pH 复合玻璃电极，磁力搅拌器，磁子。

（2）**试剂**。邻苯二甲酸氢钾标准缓冲溶液（pH＝4.00），磷酸二氢钾和磷酸氢二钠标准缓冲溶液（pH＝6.86），硼砂标准缓冲溶液（pH＝9.18），pH 未知的试液。这些标准试剂是商品化的袋装试剂，按要求配制各标准缓冲溶液即可。

实验步骤

（1）**pHS-3C 酸度计的标定（两点标定法）**。对于精密级的 pH 计，除了设有"定位"和"温度"调节外，还设有电极"斜率"调节，可以用两种标准缓冲液进行校准。一般先以 pH＝6.86 的缓冲溶液进行"定位"校准，然后根据测试溶液的酸碱情况，选用 pH＝4.00（酸性）或 pH＝9.18（碱性）的缓冲溶液进行"斜率"校正。

① 拔出电极保护帽，将电极用去离子水洗净，滤纸吸干后，浸入 pH＝6.86 的标准溶液中，用温度计测出溶液温度值，在仪器上设置此温度。待示值稳定后，按"定位"键，再按"确认"键，仪器自动识别当前溶液的 pH 值。

② 将电极用水冲洗干净，滤纸吸干后，浸入第二种标准溶液（根据测试溶液的酸碱情况选择），设置温度，按"斜率"键，再按"确认"键。仪器自动识别当前温度下标准溶液的 pH 值。

（2）**未知 pH 试液的测定**。经过标定的仪器可测量未知溶液 pH 值，不得再按斜率键和定位键。当被测溶液与标定溶液温度相同时，用去离子水清洗电极，滤纸吸干，将电极插入未知试液中，待显示屏上数据稳定后读出溶液的 pH 值。

（3）**pH 玻璃电极响应斜率的测定**。把 pH 功能键切换到 mV 档，将电极插入 pH＝4.00 的标准缓冲溶液中，在显示屏上读出溶液的 mV 值，再依次测定 pH＝6.86、pH＝9.18 标准缓冲溶液的 mV 值。

数据处理

（1）记录所测未知试液的 pH 值。

（2）以 E 对 pH 作图，求出直线斜率，该斜率即为该玻璃电极的响应斜率。若偏离 59mV/pH（25℃）太多，则该电极不能使用。

注意事项

玻璃电极易碎，不能将电极触碰烧杯底部，也不允许搅拌磁子撞击它。

思考题

（1）测定 pH 时，为什么要选用与待测溶液的 pH 相近的标准缓冲溶液来定位？

（2）为什么普通的毫伏计不能用于测量 pH？

实验 22　氟离子选择性电极法测定牙膏中总氟含量
（见视频 5）

实验目的

（1）掌握牙膏中氟离子含量的测定方法。
（2）了解总离子强度调节缓冲溶液的意义和作用。

视频 5

实验原理

氟是最活泼的非金属元素，也是人体必不可少的微量元素之一。适量氟对人体有益，摄入量过低会产生龋齿，但是摄入量长期超过正常需要，将导致地方性氟病。测定氟离子常用的方法之一是氟离子选择性电极法。它属于电化学分析中的电位分析法。该方法具有电极结构简单牢固、灵敏度高、响应速度快、能克服色泽干扰、精度高等优点，而且便于携带、操作简单，因而被广泛应用。

氟离子选择性电极（fluoride ion selective electrode，FISE）是晶体均相膜电极的一种，由 LaF_3 单晶制成，能对氟离子进行特异性识别，是用电位法测量溶液中氟离子活度的指示电极。该电极的电极电势（E_F）为

$$E_F = K - \frac{RT}{nF} \ln a_{F^-}$$

当氟离子选择性电极和参比电极组成电池后，仪器的信号即电池的电势（$E_{电池}$）与 $\ln a_{F^-}$ 成线性相关。若在测定溶液中加入适量的离子强度调节剂使离子强度保持不变，则活度系数为一常数，此时离子活度可由浓度代替，即 $E_{电池}$ 与 lg［F^-］呈线性关系。作 $E_{电池}$-lg［F^-］标准曲线，根据试样的 $E_{电池}$ 可求得氟离子浓度。在电位分析中，通常采用加入总离子强度调节缓冲溶液（TISAB）的方法来控制溶液的总离子强度。

仪器与试剂

（1）**仪器**。氟离子选择性电极，饱和甘汞电极，酸度计，电磁搅拌器。

（2）**试剂**

① 1.0×10^{-3} mol/L F^- 标准储备液。② TISAB：称取 NaCl 58g，柠檬酸钠 $Na_3C_6H_5O_7 \cdot 2H_2O$ 12g，取冰醋 57mL，溶于 500mL 水中搅拌溶解。缓缓加入 6mol/L NaOH 溶液，调节 pH 为 5.5～6.5，冷却后转移至 1000mL 容量瓶中，加水稀释至刻度线，摇匀后储存于聚乙烯瓶中。③含氟牙膏。

实验步骤

（1）**试样预处理**。准确称取含氟牙膏 0.9～1.4g 置于塑料小烧杯中，加入适量二次水充分搅拌后超声约 20min。将溶液全部转移到 100mL 容量瓶，加入 10mL TISAB 溶液，定容备用。

（2）**开机准备**。仪器预热 20min，将氟离子选择性电极、饱和甘汞电极分别与酸度计相连接，将两电极插入蒸馏水中，开动搅拌器，反复清洗电极至空白电位（－300mV）。

（3）标准曲线的制作。分别取 1.0×10^{-3} mol/L F^- 标准溶液 0.5mL、1.00mL、2.50mL、5.00mL、10.00mL 于 5 个 100mL 容量瓶中，加入 10mL TISAB 溶液，用去离子水稀释至刻度。将系列标准溶液由低浓度到高浓度依次转入干的塑料杯中，放入搅拌子，电极插入被测试液，开动搅拌器 5~8min 后停止搅拌，读取平衡电位。

（4）牙膏中含氟量的测定。将处理好的牙膏试样溶液转入干燥的塑料杯中，测 E 值。

数据处理

（1）以 E 为纵坐标，lg[F^-] 为横坐标，绘制标准曲线，得到线性回归方程。
（2）将牙膏试样电位值代入线性回归方程计算 [F^-]，并计算牙膏样中氟的含量。

思考题

（1）本实验中加入总离子强度调节缓冲溶液的目的是什么？
（2）为什么要把氟电极洗至一定的电位？

实验 23　氯离子选择性电极对溴离子选择性系数的测定

实验目的

（1）了解离子选择性电极选择性系数测定的原理和方法。
（2）掌握混合溶液法测定离子选择性电极选择性系数的实验技术。

实验原理

离子选择性电极是对特定的离子有选择性响应的电极。但这种选择性并非绝对专一，溶液中共存的其他离子对电极的电位也可能会产生一定的贡献。离子选择性电极对响应离子和其他离子的响应差异可以用电位选择性系数（selectivity coefficient）来定量表征。若测定离子为 i，j 为干扰离子，n 和 m 分别为被测离子和干扰离子的电荷数，则 $K_{i,j}$ 就是该电极的电位选择性系数。可用 Nicolsky 方程描述共存离子对电位的贡献：

$$E = K \pm \frac{2.303RT}{nF} \lg(a_i + K_{i,j} a_j^{n/m})$$

从上式可以看出，电位选择性系数越小，干扰离子 j 的干扰越小。测定 $K_{i,j}$ 的方法可以用分别溶液法或混合溶液法测定，本实验采用混合溶液法测定 $K_{i,j}$。实验时，配制一系列含有固定活度的干扰离子（j）和不同活度的被测离子（i）的标准溶液，分别测量相应的电位值 E，绘成 E-lga_i 的曲线。

当 $a_i > a_j$ 时，电极对 i 离子呈能斯特响应，此时干扰离子的影响可以忽略不计。若 i，j 离子均为一价阴离子（例如本实验），则标准曲线中的直线部分的能斯特方程为：

$$E_1 = K_1 - \frac{2.303RT}{nF} \lg a_i$$

当 $a_i < a_j$ 时，标准曲线形成水平，电极对 i 离子的响应可以忽略，电位值完全由 j 离子决定，则

$$E_2 = K_2 - \frac{2.303RT}{nF} \lg(K_{i,j} a_j^{n/m})$$

假定 $K_1 = K_2$，且两斜率相同，在两直线的交点处 $E_1 = E_2$，可以得出下述公式：

$$K_{i,j} = a_i / a_j^{n/m}$$

因此可以求得 $K_{i,j}$ 值，这一方法也称为固定干扰法。本实验以 Br^- 为干扰离子，测定氯离子选择电极的选择性系数 K_{Cl^-, Br^-}。即

$$K_{Cl^-, Br^-} = a_{Cl^-} / a_{Br^-}$$

本实验的测量体系由氯离子选择性电极、参比电极和试液组成。氯离子选择性电极的敏感膜由 Ag_2S-$AgCl$ 粉末混合压片制成。它是无内参比溶液的全固态型电极，电荷由膜内电荷数最少、半径最小的 Ag^+ 传导。由于饱和甘汞电极内的 Cl^- 可通过多孔陶瓷芯向溶液中扩散，影响 Cl^- 的测定，所以应该使用双盐桥饱和甘汞电极。

仪器与试剂

(1) **仪器**。pHS-3C 型酸度计，磁力搅拌器，氯离子选择性电极和双盐桥饱和甘汞电极。

(2) **试剂**

① 0.1000mol/L NaCl 标准溶液：准确称取 1.464g 经 110℃烘干的分析纯 NaCl 于 50mL 烧杯中，用水溶解后，转移至 250mL 容量瓶中，稀释至刻度。②0.1000mol/L NaBr 标准溶液：准确称取分析纯 NaBr 2.573g 于 50mL 烧杯中，用水溶解后，转移到 250mL 容量瓶中，用水稀释至刻度。③1.0mol/L KNO_3 作为离子强度调节剂，用 HNO_3 调节到 pH2.5 左右。

实验步骤

(1) **开机准备**

① 打开 pHS-3C 型酸度计的电源开关，按 pH/mV 功能键调整测量模式为"mV"模式。

② 检查双盐桥饱和甘汞电极是否充满 KCl 溶液，若未充满应补充饱和 KCl 溶液，并排除其中的气泡。于盐桥套管中放置 KNO_3 溶液。

③ 将氯离子选择性电极和甘汞电极连接在 pHS-3C 型酸度计上，把电极浸入蒸馏水中，开动搅拌器，将电极洗至空白电位。

(2) **溶液配制**。准确吸取适量的氯离子标准溶液于 50mL 容量瓶中，以配制 1.00×10^{-4} mol/L，1.00×10^{-3} mol/L，5.00×10^{-3} mol/L，1.00×10^{-2} mol/L，5.00×10^{-2} mol/L 和 1.00×10^{-1} mol/L NaCl 的系列标准溶液，并在上述溶液中各加入 5.00mL 1.00×10^{-2} mol/L Br^- 标准溶液，15mL 1.0mol/L KNO_3 溶液，用水稀释至刻度，摇匀。

(3) **电势测量**。将系列被测溶液从低浓度至高浓度分别转入烧杯中，电极插入上述溶液，开动搅拌器，停止搅拌后测量其平衡电位值。

数据处理

以电位 E 值为纵坐标，lgc_{Cl^-} 为横坐标作图，延长曲线中两段直线部分，并从两直线交点处求得 c_{Cl^-} 的值，根据公式计算氯离子选择性电极对溴离子的电位选择性系数。

$$K_{Cl^-, Br^-} = c_{Cl^-} / c_{Br^-}$$

思考题

(1) 评价离子选择性电极的性能有哪些特性参数？

(2) 本实验中为什么要选用双盐桥饱和甘汞电极作参比？

(3) 可否用电位选择性系数来校正氯离子响应的电位值？为什么？

实验 24　电位滴定法测定阿司匹林片剂中乙酰水杨酸的含量

实验目的

(1) 掌握电位滴定法在酸碱滴定中的应用。

(2) 熟悉采用作图法判断电位滴定终点。

实验原理

电位滴定法是以仪器代替人的眼睛，利用指示电极电位的突跃来确定滴定终点的一种分析方法。进行电位滴定时，选用适当的指示电极和参比电极与被测溶液组成一个工作电池，随着滴定剂的加入，由于发生化学反应，被测离子的浓度不断发生变化，因而指示电极的电位随之变化。在滴定终点附近，被测离子浓度发生突变，引起电极电位的突跃，因此，根据电极电位的突跃可确定滴定终点，而不需要使用化学指示剂确定终点。

理论上，任意一个酸碱滴定都可以用电位滴定法来确定终点。滴定时可在被分析的样品溶液中浸入两个已经校准的电极，其中一个是 pH 指示电极，另一个是参比电极。也可直接使用复合电极。

使用 pH 计进行酸碱电位滴定，记录滴定剂体积 (V) 和相应的 pH，按 pH-V，ΔpH/ΔV-V 或 Δ^2pH/ΔV^2-V 作图法确定终点，从而计算样品溶液的浓度。

本实验将阿司匹林片溶解于 95% 的乙醇溶液中，用 NaOH 标准溶液进行电位滴定，记录不同时段消耗的 NaOH 溶液的体积 V 和溶液的 pH 值，通过作图法测定样品中乙酰水杨酸的含量。

仪器与试剂

(1) **仪器**。pH 计 (pHS-3C 型)，复合电极，电磁搅拌器，搅拌子。

(2) **试剂**。邻苯二甲酸氢钾标准缓冲溶液 (0.05mol/L，pH = 4.00)，KH_2PO_4 和 Na_2HPO_4 标准缓冲溶液 (pH = 6.85)，95% 乙醇溶液，蒸馏水，NaOH 标准溶液 (0.1mol/L)，邻苯二甲酸氢钾 (KHP，AR)。

实验步骤

(1) **校准 pH 计**。分别用 pH = 4.00 和 pH = 6.86 的标准缓冲溶液校准 pH 计。校准步骤参照实验 21 的步骤 (1)。

(2) **NaOH 标准溶液的标定**。以 KHP 为基准物质，测定 NaOH 标准溶液的准确浓度。

(3) **电位滴定**

① 取阿司匹林片 10 片，研细，取约 0.6g，精密称定，置于 150mL 的烧杯中。加入

15mL 95％的乙醇溶液至溶解。

② 先每次滴加约 1mL NaOH 标准溶液至 pH 达 4.50 左右，再每次滴加约 0.2mL NaOH 至 pH 达 5.00 左右，每加入一定体积的 NaOH 溶液读一次 pH，在计量点前后每加入半滴或 1 滴 NaOH（0.02～0.04mL），记录一次 pH（这时 pH 变化很快）。再继续滴定至计量点后至 pH 变化小于 0.10。

数据处理

（1）记录不同时段加入的 NaOH 体积和溶液 pH 值，数据填入表 3-7。

表 3-7　不同时段加入的 NaOH 体积和溶液 pH 值

V_{NaOH}/mL	pH	ΔpH	ΔV	ΔpH/ΔV	$\bar{V}$	$\Delta^2 pH/\Delta V^2$

（2）绘制 pH-V 滴定曲线（Y 轴：被测样品 pH；X 轴：NaOH 标准液体积）。

（3）阿司匹林中乙酰水杨酸含量计算（$M_{阿司匹林}$＝180.2g/mol；$V_{NaOH,ep}$：终点时消耗的 NaOH 体积）：

$$\omega = \frac{c_{NaOH} \times V_{NaOH,ep} \times M_{阿司匹林}}{m_{阿司匹林}} \times 100\%$$

思考题

（1）如何根据 pH-V、ΔpH/ΔV-V 以及 $\Delta^2 pH/\Delta V^2$-V 作图法确定终点？

（2）试讨论本实验的误差来源。

实验 25　自动电位滴定仪测定弱酸浓度及其解离常数

实验目的

（1）了解自动电位滴定仪的主要构成部件及其功能。

（2）学习自动电位滴定仪的使用方法。

（3）学会用自动电位滴定法测定醋酸溶液的浓度及离解常数。

实验原理

自动电位滴定仪由一个与滴定管相连的吸液器，一个搅拌器，两支电极，一个滴定杯组成。吸液器用来盛装滴定剂溶液或补充滴定剂溶液，滴定杯用于盛放被测溶液。本实验用

NaOH 标准溶液滴定醋酸溶液，玻璃电极为指示电极，饱和甘汞电极为参比电极。在 25℃ 时该电池的电势为：

$$E = K + 0.05916\text{pH}$$

溶液的 pH 与电动势具有线性关系，E 值的变化可反映出溶液 pH 的变化，根据滴定曲线图可求出滴定终点，即滴定曲线图中突跃部分的中点所对应的体积为滴定终点。

乙酸（以 HAc 表示）是弱电解质，当 $[\text{Ac}^-] = [\text{Hac}]$ 时，$K_a = [\text{H}^+]$，即乙酸离解常数 K_a 等于溶液中氢离子的浓度。若达到滴定终点时所消耗的 NaOH 体积为 V_{ep}，当消耗的 NaOH 的体积为 V_{ep} 的 1/2 时，所对应的氢离子浓度等于 K_a。

仪器与试剂

(1) **仪器**。888Titrando 型自动电位滴定仪，玻璃电极，饱和甘汞电极，滴定杯，25mL 移液管。

(2) **试剂**。0.1mol/L NaOH 溶液，0.05mol/L CH_3COOH 溶液，邻苯二甲酸氢钾（KHP，AR）。

实验步骤

(1) **实验准备**。接通电源，预热仪器，按仪器操作规程将仪器调试好。

(2) **NaOH 溶液浓度的标定**。准确称取 KHP 2.5g 于烧杯中，加水溶解，转移到 250mL 的容量瓶中，定容，摇匀，用移液管准确移取 KHP 溶液 25.00mL 于滴定杯中，开动自动电位滴定仪。以滴定曲线突跃部分的中点所对应的体积为滴定终点，记下体积，取下滴定杯，用蒸馏水将滴定管和电极冲洗干净，重复三次，计算 NaOH 溶液的浓度。

(3) **醋酸溶液浓度的测定**。用移液管准确移取醋酸溶液 25.00mL 于滴定杯中，开动自动电位滴定仪，仔细观察，以滴定曲线突跃部分的中点所对应的体积为滴定终点，记下体积，取下滴定杯，用蒸馏水将滴定管和电极冲洗干净，重复三次，计算醋酸的浓度。

(4) **关机**。切断电源，清洗电极和滴定管，将电极用滤纸擦干后放回电极盒内。

数据处理

将 NaOH 溶液浓度标定的有关数据填入表 3-8，HAc 浓度和解离常数测定的数据分别填入表 3-9 和表 3-10。

表 3-8　NaOH 溶液浓度的标定

项目	1	2	3	平均
KHP 质量/g				
消耗 NaOH 体积/mL				
NaOH 浓度/(mol/L)				

表 3-9　HAc 浓度的测定

项目	1	2	3	平均
HAc 体积/mL	25.00	25.00	25.00	25.00
消耗 NaOH 体积/mL				
HAc 浓度/(mol/L)				

表 3-10　HAc 解离常数的测定

项目	1	2	3	平均
$1/2V_{ep}$ 的 NaOH 体积/mL				
$1/2V_{ep}$ 的 pH				
K_a				

注意事项

甘汞电极、玻璃电极使用中要小心轻放，防止打碎。玻璃电极用后要用蒸馏水浸泡。

思考题

（1）简述化学滴定法与电位滴定法的不同之处。

（2）列出电位滴定仪的使用步骤。

实验 26　亚铁氰化物/铁氰化物氧化还原电对的循环伏安法研究

实验目的

（1）学习电化学工作站的使用。

（2）学习抛光固体电极表面的处理方法。

（3）学会判断电极过程的可逆性。

实验原理

循环伏安法是研究电活性物质的一种常用的电分析技术。它是在工作电极上施加三角形的脉冲电压，该电压随时间线性变化，当达到设定的终止电压后，又反向扫描至初始电压。此时便完成了一个电压变化的循环。得到的响应电流随外加电压变化的关系图就是循环伏安图。

如果一个氧化还原电对的半反应能在工作电极上快速完成电子转移，这样的氧化还原电对称为电化学可逆电对。亚铁氰化钾/铁氰化钾 $\{[Fe(CN)_6]^{4-}/[Fe(CN)_6]^{3-}\}$ 是典型的可逆氧化还原体系。当电压正向扫描时，$[Fe(CN)_6]^{4-}$ 被氧化，此时的阳极电流由以下反应产生：

$$[Fe(CN)_6]^{4-} \longrightarrow [Fe(CN)_6]^{3-} + e^-$$

电极作为氧化剂，氧化电流逐渐增加。当电极表面 $Fe(CN)_6^{4-}$ 的浓度逐渐减小直到耗尽时，电流达到峰值，随后电流逐渐减小。当扫描方向切换为负向时，此时的电势仍足以氧化 $[Fe(CN)_6]^{4-}$，因此电极上也会有阳极电流继续产生。当电势扫描到足够负时，此时靠近电极表面形成的 $[Fe(CN)_6]^{3-}$ 将被还原：

$$[Fe(CN)_6]^{3-} + e^- \longrightarrow [Fe(CN)_6]^{4-}$$

当靠近电极表面的 $Fe(CN)_6^{3-}$ 耗尽时，此时阴极电流达到峰值然后衰减。由循环伏安图上可以得到阳极峰值电流 i_{pa}、阴极峰值电流 i_{pc}、阳极峰值电位 E_{pa} 和阴极峰电位 E_{pc}。

可逆的氧化还原体系，氧化峰电流和还原峰电流比值 $i_{pa}/i_{pc}=1$。氧化峰电位与还原峰电位差 $\Delta E = E_{pa} - E_{pc} \approx 0.059/n$（$n$ 为可逆电对的氧化还原反应中涉及的电子转移数）。如果电活性物质可逆性差，如不可逆过程，则峰电位的差值增大，i_{pa} 和 i_{pc} 大小也不同。由此可判断电极过程的可逆性。

对于扩散控制的可逆体系，峰电流与物质浓度符合 Randles-Sevcik 方程：

$$i_p = 2.69 \times 10^8 n^{3/2} A D^{1/2} v^{1/2} c$$

式中，i_p 为峰电流；n 为电子转移数；A 是电极面积；D 是扩散系数，m^2/s；c 为被测物质浓度（mol/L）；v 为扫描速率（V/s）。

本实验配制不同浓度的 $K_3Fe(CN)_6$/ $K_4Fe(CN)_6$ 标准溶液，考察循环伏安法中峰电流与物质浓度和扫描速率的关系。同时，考察可逆电对的电化学行为。

仪器与试剂

(1) **仪器**。CHI 660E 电化学工作站（上海辰华），金电极，铂丝电极，饱和甘汞电极，电极抛光材料。

(2) **试剂**。0.30mol/L $K_3Fe(CN)_6$/ $K_4Fe(CN)_6$ 标准溶液（含 1.0mol/L 的氯化钾）。

实验步骤

(1) **系列标准溶液的配制**。分别取 3.00mL、5.00mL、6.00mL、7.00mL 及 8.00mL $K_3Fe(CN)_6$/ $K_4Fe(CN)_6$ 标准溶液（0.30mol/L，含 1.0mol/L 的氯化钾）于 50mL 容量瓶中，再加入 1.0mol/L 的氯化钾溶液至刻度定容，摇匀，备用。

(2) **工作电极的预处理**。工作电极用三氧化二铝（Al_2O_3）粉末（粒径 $0.05\mu m$）将电极表面抛光，然后用蒸馏水清洗，干燥。

(3) **仪器参数设置**

① 打开 CHI 660E 电化学工作站和计算机的电源预热 10min。

② 在电解池中放入配制好的标准溶液，插入电极，以新处理的金电极为工作电极，铂丝电极为对电极，饱和甘汞电极为参比电极，接好测量电路（红色夹子接 Pt 辅助电极、绿色接金电极、白色接甘汞参比电极）。

③ 打开 CHI 660E 的 Setup（设置）下拉菜单，在 Technique（技术）项选择 Cyclic Voltammetry（循环伏安法），在 Parameters（参数）项内设定如下参数（以上参数按照给定的体系进行适当调整）：

初始电位（Init E）——$-0.2V$

最高电位（High E）——$+0.6V$

最低电位（Low E）——$-0.2V$

终止电位（Final E）——$+0.6V$

扫描速率（Scan Rate）——0.1V/s

采样间隔（Sample Interval）——0.001V/s

初始电位下的极化时间或停止时间（Quiet Time）——5s

电流灵敏度（Sensitivity）——0.001

完成上述各项参数设定，再仔细检查一遍无误后，点击"OK"键。然后点击工具栏中的运行键，此时仪器开始循环伏安扫描，屏幕上即时显示电流对电位的曲线。测量完成后，保存谱图。

④ 不同浓度时的循环伏安图。将配制好的 $K_3Fe(CN)_6$ 标准溶液分别倒入电解池中，插

入无水分的工作电极、对电极和饱和甘汞电极，以 100mV/s 的扫描速率，从 0.6V 到 -0.2V 完成扫描，分别记录各种浓度溶液的循环伏安图。

⑤ 不同扫描速率时的循环伏安图。在一定浓度的 $K_3Fe(CN)_6$ 溶液中，以不同扫描速率 50mV/s、80mV/s、100mV/s、150mV/s、200mV/s，分别记录从 0.6V 至 -0.2V 扫描的循环伏安图。

数据处理

(1) 由 $K_3Fe(CN)_6/K_4Fe(CN)_6$ 溶液的循环伏安图记录 i_{pa}、i_{pc} 和 E_{pa}、E_{pc} 值。

(2) 分别以 i_{pa} 和 i_{pc} 对 $v^{1/2}$ 作图，说明扫描速率 v 对 i_p 的影响。

(3) 说明扫描速率对 ΔE_p 的影响。

(4) 分别以 i_{pa} 和 i_{pc} 对 $K_3Fe(CN)_6$ 的浓度作图，说明浓度与峰电流的关系。

(5) 从实验结果判断 $[Fe(CN)_6]^{4-}/[Fe(CN)_6]^{3-}$ 氧化电对的可逆性。

注意事项

(1) 电极的处理，金电极在测定前用抛光粉抛光，超声清洗并用水冲洗净抛光粉才可使用。工作电极表面处理应耐心细致，否则严重影响实验结果。

(2) 为了使液相传质过程只受扩散控制，应在加入电解质和溶液处于静止下进行电解。

(3) 不同扫描之间，为使电极表面恢复初始状态，应将电极提起后再放入溶液中，或将溶液搅拌，等溶液静止后再扫描。

(4) 避免电极夹头互碰导致仪器短路。

思考题

(1) 实验前电极表面为什么要处理干净？

(2) 扫描过程中溶液为什么要保持静止？

实验 27　微分脉冲伏安法测定果汁饮料中的维生素 C

实验目的

(1) 掌握微分脉冲伏安法的基本原理。

(2) 掌握微分脉冲伏安法测定维生素 C 含量的方法。

实验原理

微分脉冲伏安法是一种高灵敏度的伏安分析技术。它是在缓慢变化的直流电压上叠加一个恒振幅的脉冲电压，脉冲振幅（高度）一般十至几十毫伏，持续时间 40~60ms，记录脉冲结束前一瞬间的电流与加脉冲前一瞬间的电流之差。由于采用了两次电流取样的方法，因而能很好地扣除因直流电压引起的背景电流，分析灵敏度比循环伏安法高。微分脉冲伏安法的峰电流与被测物质的浓度成正比，可用于物质的定量分析。

维生素 C 又名抗坏血酸（Vc），是一种人体所必需的化学物质。它可在电极表面失去电子被氧化，产生氧化电流。通过标准曲线法可测定维生素 C 果汁饮料或维生素片剂等实际样品中维生素 C 的含量。

仪器与试剂

(1) 仪器。 CHI 660E 电化学工作站（上海辰华），pH 酸度计，超声波清洗器，饱和甘汞电极，铂电极，玻碳电极，电极抛光材料，100mL 吸量管，10mL 移液管，烧杯，玻璃棒，滴管若干。

(2) 试剂。 维生素 C（分析纯，$C_6H_8O_6$，$M_r = 176g/mol$），维生素 C 果汁饮料，去离子水，无水乙醇，KCl，KNO_3，HCl，NaOH，pH＝4.00 磷酸盐缓冲溶液，50mmol/L 的铁氰化钾溶液其中氯化钾含量 0.5mol/L，500mL 1∶1 HNO_3 溶液。

实验步骤

(1) 玻碳电极预处理。 将玻碳电极用少量 α-Al_2O_3 粉在麂皮上抛光，蒸馏水冲洗表面污物，如有需要可依次用 1∶1 乙醇、1∶1 HNO_3 和蒸馏水进行超声清洗，每次 2～3min。电极抛光后移取 15mL 铁氰化钾溶液到电解池中，插入干净的电极系统。起始电位 0.6V，终止电位 -0.2V，以 50mV/s 的扫描速率进行测量，峰电位差若在 100mV 内，说明该电极可进行下面相关实验。

(2) 标准曲线的绘制

① 准确称取维生素 C 标准品适量（约 0.22g），溶于 pH＝4.00 的磷酸盐溶液，由 50mL 棕色容量瓶定容备用（约 25mmol/L）。

② 分别准确移取 0.50mL、1.00mL、1.50mL、2.00mL、2.50mL 维生素 C 标准溶液到电解池，加适量的磷酸盐缓冲溶液（总体积为 10mL）。在选定的最佳实验条件下，以处理好的玻碳电极为工作电极、饱和甘汞电极为参比电极、铂电极为辅助电极，用微分脉冲伏安法扫描，记录 *i-E* 曲线。每次扫描结束后取出电极用二次水冲洗，滤纸吸干后再进行下一次测定，这样可保持修饰电极良好的稳定性和重现性。绘制峰电流与维生素 C 浓度的标准曲线。

(3) 样品测定。 移取果汁样品 10mL 至 50mL 容量瓶中，用磷酸盐缓冲溶液定容至刻度，取 10mL 进行测定，平行测定三次，得到微分脉冲伏安曲线图。

数据处理

(1) 以峰电流对维生素 C 标准溶液浓度作图得到校准曲线和线性回归方程。

(2) 计算维生素 C 果汁饮料中维生素 C 的含量。

思考题

(1) 微分脉冲伏安法为何能达到较高的灵敏度？

(2) 微分脉冲伏安图为什么呈现峰形？

实验 28　阳极溶出伏安法测定水样中微量铅和镉

实验目的

(1) 掌握阳极溶出伏安法的基本原理。

（2）学习电化学工作站阳极溶出伏安功能的使用方法。

（3）掌握使用单次标准加入法进行定量分析。

实验原理

阳极溶出伏安法（anodic stripping voltammetry，ASV）的测定包含两个基本过程。首先，将工作电极控制在一定的负电位条件下使被测物质富集在电极表面，然后施加一定变化的电压于工作电极上，使被富集的物质溶出，根据溶出峰电流的大小测定被测物质的含量。

ASV 是用于测定水中（如河流，湖泊，饮用水源等）金属离子污染物的最灵敏，便捷和性价比高的分析方法之一。它可以同时进行多元素的痕量分析，如 Pb，Cd，Cu 和 Zn 等。检测灵敏度可达 mg/L 甚至 μg/L 水平。

如测定水样中痕量镉和铅时，可在酸性介质中，控制电极电位为 -1.0V（vs. SCE），此时 Pb^{2+} 和 Cd^{2+} 同时富集在工作电极（自制玻碳汞膜电极）上，然后当向阳极线性扫描至 -0.1V（vs. SCE）时，可以得到两个溶出峰，铅的溶出峰电位在 -0.4V（vs. SCE）左右，镉的溶出峰电位在 -0.6V（vs. SCE）左右。电流与溶液中 Pb^{2+} 和 Cd^{2+} 浓度成正比，可以分别用于铅和镉的定量分析。

定量测定可采用标准曲线法或标准加入法。标准加入法的计算公式如下：

$$c_x = \frac{c_s V_s i_x}{i(V_x + V_s) - i_x V_s}$$

式中，c_x、V_x 和 i_x 分别为试样的浓度、体积和溶出峰的峰电流；c_s 和 V_s 分别为加入的标准溶液的浓度和体积；i 为加入标准溶液后测得的溶出峰的峰电流值。

仪器与试剂

（1）**仪器**。CHI 660E 电化学工作站，玻碳电极，铂丝电极，饱和甘汞电极，25mL 电解池，磁力搅拌器，搅拌磁子，氮气钢瓶。

（2）**试剂**。1.000mg/mL Pb^{2+} 标准溶液，1.000mg/mL Cd^{2+} 标准溶液，0.02mol/L $HgSO_4$ 溶液，2.0mol/L HAc-NaAc 溶液（pH=5.0）。

实验步骤

（1）**玻碳电极的制备**。抛光玻碳电极，在电解池中加入 10mL 蒸馏水和 100μL $HgSO_4$ 溶液，将三电极系统插入溶液并与电化学工作站连接。控制电极电位 -1.0V，通氮气搅拌下，电镀 5min 即可制得玻璃碳汞膜电极。

（2）**铅和镉峰电位的测量**

① 调节电化学工作站参数，选择单扫描模式，起始电位 -1.2V，终止电位 $+0.5$V，扫描速率 100mV/s。

② 在电解池中加入 10.00mL 蒸馏水和 1.00mL HAc-NaAc 溶液，通氮气除氧 10min，插入三电极系统，打开搅拌器，电解富集 60s（富集电位 -1.2 V）。关闭搅拌器，停止富集，静置 30s 后，开始扫描，记录空白溶出曲线。

③ 在上述空白溶液中加入 20.0μL 1.000mg/mL Pb^{2+} 标准溶液和 200μL 1.000mg/mL Cd^{2+} 标准溶液，重复②的操作，记录溶出曲线。测量结束后，将三电极系统置于 $+0.1$V 下清洗 30s。

④ 增加 Pb^{2+} 和 Cd^{2+} 的量，改变实验条件如富集时间、扫描速率、富集电位等，观察溶出曲线的变化，确定铅和镉的峰电位值，以此作为定性分析的依据。

（3）定量测定

① 在电解池中加入 10.00mL 待测水样和 1.00mL HAc-NaAc 溶液，按步骤（2）中的操作，记录溶出伏安曲线，并重复 2 次。

② 在上述电解池中加入一定量的铅和镉标准溶液（加入量视水样中待测离子的含量而定），再次记录伏安曲线，并重复 2 次。

数据处理

由伏安曲线上加入标准溶液前后两次峰高（峰电流）的值，按标准加入法计算公式计算水样中铅和镉的含量。

注意事项

（1）如果所用试剂空白值较大，计算含量时需扣除空白值，以免产生较大误差。

（2）所用测试液中均含有汞，必须倒入指定的回收瓶中，禁止倒入水槽，以免造成环境污染。

思考题

（1）为什么阳极溶出伏安法有较高的灵敏度？

（2）影响阳极溶出伏安法的因素有哪些？应如何控制？

实验 29　库仑滴定法测定维生素 C 药片中的抗坏血酸含量（见视频 6）

实验目的

（1）熟悉库仑仪的使用方法和有关操作技术。

（2）学习和掌握库仑滴定法测定抗坏血酸（维生素 C）的基本原理。

视频 6

实验原理

维生素 C 是维持人体健康的最重要的维生素之一。目前测定维生素 C 含量的方法有碘量法、紫外分光光度法、伏安法、红外光谱法、伏安分析法以及库仑滴定法等。其中，库仑滴定法的优点是不需要配制及标定标准溶液，以电解液直接进行滴定，分析结果通过精确测定电量或电位而获得，因而具有灵敏度高、精密度好和准确性高的特点。

库仑滴定是通过由电解（恒电流或恒电位）产生的滴定剂，在电解池中与被测定物质定量反应来测定该物质的一种分析方法。若电解的效率为 100%，电生滴定剂与被测物质的反应是完全的，而且有灵敏的确定终点的方法，那么，所消耗的电量与被测物质的量成正比。

根据法拉第定律可进行定量计算。计算式如下：

$$m = QM/nF$$

式中，m 是被滴定物质的质量；Q 为电极反应所消耗的电量；M 是被测物质的摩尔质量；F 为法拉第常数（96487 C/mol）；n 是电极反应的电子转移数。

本实验采用配有 4 支 Pt 电极的 KLT-1 型通用库仑仪（2 支发生电解反应，另 2 支发生终点指示反应），用恒电流电解过量的 KBr 酸性溶液，产生的电解反应如下：

$$阳极 \qquad 2Br^- === 2e^- + Br_2$$

$$阴极 \quad 2H^+ + 2e^- === H_2\uparrow$$

电解产生的 Br_2 与抗坏血酸发生下述反应：

$$O=C-O-CH-C-CH_2OH + Br_2 \longrightarrow O=C-O-CH-C-CH_2OH + 2HBr$$

当抗坏血酸消耗完时，Br_2 的浓度突然升高，表明滴定反应已经结束。Br_2 浓度的升高导致两个终点 Pt 指示电极之间产生电流。在滴定反应完全前，在这两个电极之间施加约 150mV 的电压不足以电解任何溶质，因此只有 $<1\mu A$ 的微小电流流过微安表，微安表的指针不会偏转。在化学计量点时，抗坏血酸反应完全，$[Br_2]$ 突然增加，微安表的指针由于下述可逆反应产生的电流而发生偏转：

$$检测阳极 \qquad 2Br^- === 2e^- + Br_2$$

$$检测阴极 \quad Br_2 + 2e^- === 2Br^-$$

被测物质的含量可通过滴定反应消耗的电量及被测物质转移的电子数计算而得。

仪器与试剂

(1) 仪器。KLT-1 型通用库仑仪，磁力搅拌器，电解池（双铂工作电极，双铂终点指示电极），移液管 1mL，量筒 100mL。

(2) 试剂

① 电解液：1：2 HAc 与 0.5mol/L KBr 溶液等体积混合。②0.5g/L 抗坏血酸标准溶液：准确称取标准品 0.05g，溶于去离子水定容至 100mL。③样品：准确称取一片维生素 C 药片于小烧杯中，用少量蒸馏水浸泡片刻，用玻璃棒小心捣碎，在超声波清洗器中助溶。药片溶解后（药片中有少量辅料不溶），把溶液连同残渣全部转移到 50mL 容量瓶中，用蒸馏水定容至刻度。

实验步骤

(1) 按仪器使用说明打开电源开关，预热 20～30min。

(2) 量取 KBr 电解液 50mL 于电解池中，放入搅拌子。用滴管取电解液滴入工作阴极套管内，使其高出外部液面。将清洁的电极插入溶液，把电解池放在磁力搅拌器上，用夹子固定好，开启搅拌器，调节适当的转速。

(3) 把量程选择到 5mA，电流微调（顺时针）先放在最大位置，工作选择按"电流"键（指示电极应夹在两个铂片指示电极上）。滴定终点曲线变化选择"上升"键，然后按下"启动"键，再按下"极化电位"键，调节极化电位旋钮，使表头指示极化电位约为 150mV，释放"极化电位"键。按一下"电解"按钮，红灯灭，将"工作/停止"开关置于"工作"，进行预电解，电解到终点时表针迅速向右偏转，红灯亮，电解终止。

（4）释放"启动"键，取 1.00mL 抗坏血酸标准溶液于电解池中，插好电极，掀下"启动"键，按一下"电解"键进行电解，记下计量点时所消耗的电量（mC）。再重复测定样品溶液两次（注意：取标准溶液和样品溶液各用一只移液管）。

（5）用样品溶液按步骤（4）测两次。

（6）实验完毕后应洗净电解池及电极，并注入蒸馏水。

数据处理

（1）电流效率的计算

$$\eta = Q_{标}/Q_{测} = [(m/M)nF]/Q_{测} = (CV)_{标} nF/(1000Q_{测})$$

（2）记录数据于表 3-11，并计算药片中维生素 C 含量。

$$m = M\eta Q_{测}/(nFm_{样})$$

表 3-11　维生素 C 药片中抗坏血酸的分析结果

测量次数 n	维生素 C 药片质量/g	消耗电量/mC	药片中抗坏血酸含量/(mg/g)		
			单次值	平均值	相对平均偏差/%

思考题

（1）电解液中加入 KBr 和冰醋酸的作用是什么？

（2）所用的 KBr 如果被空气中的 O_2 氧化，将对测定结果产生什么影响？

（3）电解过程中，阴极不断出现 H_2 会对电解液的 pH 值有何影响？

（4）为何电解电极的阴极要置于保护套中，而终点指示电极则不需要？

实验 30　混合酸的电导滴定

实验目的

（1）掌握电导滴定测定混合酸浓度的基本原理和实验方法。

（2）掌握 DDS-11A 型电导率仪的结构、性能和使用方法。

实验原理

在一定温度下，电解质溶液的电导率与溶液中的离子组成和浓度有关，而滴定过程中溶液的离子组成和浓度都在不断变化，因此可以利用电导率的变化来指示反应终点。电导滴定法就是在加入一种反应物时连续观察反应混合物的电解电导率，并借助滴定终点前后电导率的变化来指示反应终点的一种滴定分析方法。用 NaOH 溶液滴定 HCl 和 HAc 的混合液时，HCl 首先被中和，溶液中迁移速度较大的 H^+ 被加入的 OH^- 中和而生成难电离的水以及迁移速度较小的 Na^+，反应如下：

$$HCl + NaOH \Longleftrightarrow H_2O + NaCl$$

由于 Na^+ 的摩尔电导小于 H^+ 的摩尔电导（$\lambda_{\Theta,Na^+}=50.0\times10^{-4}\Omega^{-1}$，$\lambda_{\Theta,H^+}=341.8\times10^{-4}\Omega^{-1}$），因此，在化学计量点前，随着滴定的进行，溶液的电导逐渐下降，在化学计量点后，随着过量的氢氧化钠溶液的加入，溶液的电导随之增大。

当 HCl 被完全中和后，CH_3COOH 开始被中和生成难电离的水和易被解离的 NaAc，其反应如下：

$$CH_3COOH + NaOH \Longleftrightarrow CH_3COONa + H_2O$$

在这个滴定反应中，随着醋酸转化为醋酸盐，溶液电导有所增加，当氢氧化钠过量时，由于溶液中 OH^- 迁移速率较大使电导迅速上升。

以溶液的电导为纵坐标，滴定剂氢氧化钠的体积为横坐标，可得到具有两个拐点的滴定曲线。根据拐点处氢氧化钠的体积，可计算出 HCl 和 HAc 各自的含量。

仪器与试剂

(1) 仪器。DDS-11A 型电导率仪，磁力搅拌器，DJS-2 铂黑电导电极，碱式滴定管，烧杯，移液管，吸耳球。

(2) 试剂。0.1mol/L NaOH 标准溶液，1% 酚酞指示剂，未知混合酸，邻苯二甲酸氢钾（KHP，AR）。

实验步骤

(1) 氢氧化钠标准溶液的标定。以 KHP 为基准物质，测定 NaOH 标准溶液的准确浓度。

(2) 混合酸的滴定

① 准确移取混合酸 25mL 于 250mL 烧杯中，用去离子水稀释至 100mL 左右。

② 放入搅拌子，并将电导电极插入溶液中。

③ 启动磁力搅拌器，调整至合适的搅拌速率。记录电导值。

④ 开始滴加 NaOH 滴定剂，并在每次滴加 0.5mL 后记录电导值，直到添加约 30mL 碱。

⑤ 重复步骤①到④并记录结果。

⑥ 根据消耗的氢氧化钠体积和记录的电导值，绘制滴定曲线。

⑦ 分别计算混合酸中 HCl 和 HAc 各自的浓度。

数据处理

(1) 根据氢氧化钠溶液的标定结果，利用下式计算 NaOH 标准溶液的准确浓度。

$$c_{NaOH} = \frac{m_{KHP}}{M_{KHP}V_{NaOH}} \times 1000 \ (mol/L)$$

(2) 以记录的电导值为纵坐标，消耗的滴定剂体积为横坐标，绘制滴定曲线。根据 3 条直线上的拐点，分别计算 HCl 和 HAc 的含量。

思考题

(1) 解释用氢氧化钠滴定盐酸和醋酸的电导滴定曲线有何不同？为什么？

(2) 电导滴定有哪些优缺点？

实验 31　气相色谱法测定正构烷烃的组成和含量
（见视频 7）

视频 7

实验目的

（1）学习气相色谱仪的基本结构及操作方法。

（2）掌握面积归一化法定量的方法。

实验原理

气相色谱法（gas chromatography，GC）是一种以气体为流动相的色谱分离分析技术。载气将汽化后的试样带入加热的色谱柱，并携带分离组分通过固定相，达到分离的目的。根据色谱流出曲线（色谱图）上峰的位置（保留时间）进行定性分析，根据色谱峰的面积或峰高的大小进行定量分析。有三种常用的色谱定量方法：外标法，内标法和归一化法。

本实验目的为测定混合烷烃的组成和含量，采用纯样对照法根据保留时间进行定性分析，用不带校正因子的峰面积归一化法进行定量分析。其中定量分析的依据是在一定条件下，被测物质的质量 m 与检测器的响应值成正比，即：

$$m_i = f_i A_i$$

或

$$m_i = f_i h_i$$

式中，A_i 为被测组分的峰面积；h_i 为被测组分的峰高；f_i 为校正因子。

由于组分的含量与其峰面积成正比，如果样品中的所有组分都能产生信号并获得相应的色谱峰，则可以使用以下归一化公式来计算每种组分的含量。即把试样中所有组分的含量之和按 100% 计算，并分别求出样品中各组分的峰面积和校正因子，然后依次求各组分的百分含量。

$$\omega_i = \frac{A_i f_i}{\sum A_i f_i} \times 100\%$$

如果样品中各组分的校正因子相似，则可消除校正因子，直接使用峰面积归一化进行计算。

仪器与试剂

（1）仪器。 7820A 气相色谱仪［配有氢火焰离子化检测器（FID）］，SE-54 色谱柱（30m×0.32mm×0.33μm），1μL 微量注射器。

（2）色谱条件。 载气：N_2，34mL/min；进样口温度：220℃；分流比：1∶30；FID 检测器参数：温度为 250℃，氢气流速为 30mL/min，空气流速为 300mL/min；程序升温：先升高柱温至 80℃并保持 1min，后以 25℃/min 升温至 180℃并保持 1min。

（3）试剂。 烷烃混合样（丙酮为溶剂），烷烃标准品（正庚烷、正辛烷、正壬烷、正癸烷）。

实验步骤

（1）开机和参数设置

① 打开载气（N_2），打开气相色谱仪，打开电脑；打开主界面上的柱箱、检测器、进样口加热开关。

② 建立分析方法，设定色谱条件。

③ 待仪器稳定后（即基线平稳），即可进样分析。

（2）样品分析测定

① 采用 $1\mu L$ 微量注射器分别准确吸取各烷烃标样 $0.4\mu L$，依次注入仪器，待出峰后记录各自保留时间 t_R（min）和峰面积（A）。

② 同样色谱条件下，用 $1\mu L$ 微量注射器吸取烷烃混合样 $0.4\mu L$，注入仪器，待出峰后记录其 t_R（min）和峰面积（A）。

（3）关机

① 关闭色谱工作站主界面上的柱箱、检测器、进样器加热开关。

② 关闭氢气、空气。

③ 待检测器温度降至 $50℃$ 后，关闭色谱仪。

④ 关闭载气。

数据处理

（1）根据各标准样和混合烷烃的 t_R，对样品中各峰进行归属。

（2）根据各组分的峰面积，采用归一化法依次求出各组分的百分含量。

思考题

（1）色谱定量方法有哪些？指出它们的适用范围。

（2）进样操作有哪些注意事项？

实验 32　气相色谱法测定漱口水中百里酚的含量

实验目的

（1）掌握内标法的定量原理。

（2）掌握用气相色谱法测定百里酚浓度的方法。

实验原理

百里酚是一种天然的生物杀菌剂。它具有很强的防腐性能，是漱口水的成分之一。漱口水中的百里酚经分离后使用带氢火焰离子化检测器（FID）的气相色谱仪分离、测定，采用内标法进行定量分析。

内标法是将已知量的内标物（与分析物不同）添加到待测组分中，通过比较待测组分与内标物的信号求出该组分含量的方法。该方法特别适用于样品量或仪器响应信号在每次运行之间略有不同的分析。

本实验中，在百里酚的系列标准溶液中，分别依次加入一定量的内标物十四烷烃，将含有内标物的系列标准溶液进行色谱分析。以百里酚标准溶液的峰面积和十四烷烃的峰面积的比值作为纵坐标，标准溶液的浓度作为横坐标，绘制标准曲线并得到线性回归方程。然后在同样色谱条件下，将含有内标物的漱口水进行色谱分析，由漱口水中百里酚和十四烷烃的峰面积比值代入回归方程，即可计算出漱口水中百里酚的含量。

仪器与试剂

(1) **仪器**。7820A 气相色谱仪［配有氢火焰离子化检测器（FID）］，HP-INNOWax（聚乙二醇）色谱柱（30m×0.32mm×0.25μm），1μL 微量注射器，10mL 容量瓶。

(2) **色谱条件**。载气：N_2，1.5mL/min；进样口温度：240℃；不分流进样；FID 检测器参数：温度为 260℃，氢气流量为 40mL/min，空气流量为 400mL/min；程序升温：先升高柱温至 100℃并保持 2min，后以 10℃/min 升温至 240℃并保持 5min。

(3) **试剂**。甲醇，1000mg/L 百里酚溶液，漱口水，十四烷烃。

实验步骤

(1) 绘制含内标物的标准曲线

① 从 1000mg/L 储备溶液中，分别配制至少 5 个浓度的标准溶液于 10mL 容量瓶中，浓度范围在 100～500mg/L。

② 在每个容量瓶中各加入 5μL 十四烷烃（作为内标溶液），摇匀。用甲醇稀释至刻度。

③ 分别将不同浓度的标准溶液注入 GC 系统，记录保留时间和峰面积。

(2) 样品测定

① 移取 5.00mL 的漱口水于 10mL 的容量瓶，在溶液中加入 5μL 十四烷烃并充分混合，用甲醇稀释至刻度，摇匀。

② 将稀释的漱口水注射到 GC 系统，找出色谱图中百里酚和十四烷烃的峰及其峰面积。

数据处理

(1) 计算百里酚峰面积与十四烷烃峰面积之比，绘制此系列比值对浓度的校准曲线，并得到回归方程。

(2) 计算漱口水中百里酚峰面积与十四烷烃峰面积之比。使用回归方程计算漱口水中的百里酚浓度（mg/L）。

思考题

(1) 讨论实验中的误差来源并估算所得结果的不确定性。

(2) 思考样品的色谱图中为何有更多的峰出现？

实验 33　高效液相色谱法测定水样中间硝基苯酚含量
（见视频 8）

实验目的

(1) 了解高效液相色谱仪的基本结构及操作方法。

(2) 了解高效液相色谱法的工作原理。

(3) 掌握高效液相色谱法定性和定量分析的基本方法。

视频 8

实验原理

高效液相色谱（high-performance liquid chromatography，HPLC）采用液体作为流动相，利用物质在流动相和固定相两相中吸附或分配系数的差异达到快速分离。HPLC 主要由自动进样器、高压输送系统、进样阀、色谱柱、检测和数据处理系统组成。

反相色谱（reversed-phase chromatography）是一种常用的 HPLC 技术，其中固定相是非极性或弱极性的，流动相溶剂是极性的。反相高效液相色谱法采用的固定相一般为十八烷基键合硅胶（ODS），流动相为水、甲醇、乙腈等。当混合物中各组分在两相的分配系数不同时，各组分随流动相沿色谱柱出口方向移动，就会产生差速迁移而分离。分配系数小的组分先出峰，分配系数大的组分后出峰。

间硝基苯酚是一种极其有害的环境污染物，也是环境水中的污染物之一。本实验采用反相高效液相色谱法分离测定间硝基苯酚，根据系列标准溶液的峰面积对浓度绘制工作曲线，再根据样品的峰面积得到间硝基苯酚的浓度。

仪器与试剂

（1）**仪器**。Waters 600 高效液相色谱仪，Waters 2996PDA 检测器，Empower 操作系统，100μL 微量进样器，20μL 定量环，0.45μm 滤膜，10mL 刻度管，1mL 吸量管。

（2）**色谱条件**。Symmetry C_{18} 色谱柱（150mm×3.9mm×5μm）；流动相为 70% 的甲醇＋30% 的水（体积分数），流速为 0.8mL/min，室温，检测波长 271nm。

（3）**试剂**。间硝基苯酚储备液 0.1mg/mL，水样。

实验步骤

（1）分别准确吸取间硝基苯酚储备液 0.2mL、0.4mL、0.6mL、0.8mL、1.0mL 于 10mL 刻度管中，用水稀释至刻度，摇匀。该系列标准溶液浓度分别为 2.0μg/mL、4.0μg/mL、6.0μg/mL、8.0μg/mL、10μg/mL。

（2）打开真空脱气机、泵、检测器开关。启动电脑，在电脑中编辑泵和检测器的仪器方法，并命名和保存。

（3）建立仪器方法，输入样品名称、样品号、运行时间等。

（4）使流动相通过色谱柱 5～10min，待基线稳定后开始进样。

（5）将进样阀放在装载（LOAD）位时，用微量进样器取约 60μL 浓度为 2μg/mL 的标准溶液（用 0.45μm 滤膜过滤），注入进样阀中，此时溶液装载在定量环中。

（6）将进样阀从装载（LOAD）位转向进样（INJECT）位，此时溶液随着流动相进入色谱柱进行分离，记录保留时间和峰面积。

（7）按标准溶液浓度增加的顺序，重复步骤（5）、（6）操作。

（8）样品分析按步骤（5）、（6）操作。平行测定 3 次，记录其保留时间和峰面积。

数据处理

对系列间硝基苯酚的色谱峰进行积分，以峰面积为纵坐标，浓度为横坐标绘制工作曲线并得出线性回归方程及相关系数。将水样中间硝基苯酚的峰面积代入回归方程计算间硝基苯酚的浓度。

思考题

(1) 正相色谱和反相色谱有什么相同点和不同点？

(2) 说明用反相色谱测定间硝基苯酚的原理。

(3) 高效液相色谱仪有哪些常用检测器？

(4) 流动相为什么必须用滤膜过滤？

实验 34　高效液相色谱法测定槐米中芦丁的含量

实验目的

(1) 掌握高效液相色谱法定性及定量分析的基本方法。

(2) 熟悉高效液相色谱仪的工作原理和操作方法。

实验原理

槐米系豆科植物槐树（*Sophora japonica* L.）的干燥花蕾，具有清热抗炎的功效。槐米中主要含有黄酮苷、皂苷、甾醇等成分，其中芦丁（rutin）含量最高。芦丁具有调节毛细血管壁的渗透性的作用，临床上用作毛细血管止血药，可作为高血压的辅助治疗药物。芦丁可溶于甲醇、水等溶剂，在 254nm 处有较大吸收。芦丁的结构式如图 3-7 所示。

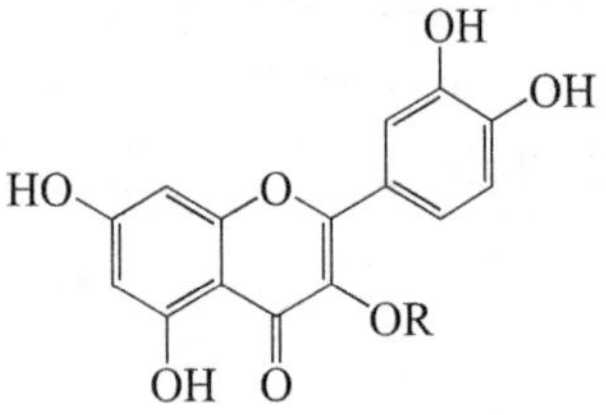

R = 芸香糖基

图 3-7　芦丁的结构式

本实验采用反相色谱法将槐米中的芦丁和其他组分分离后，进行紫外检测。在一定的实验条件下，以保留时间作为定性参数，峰面积作为定量参数，以系列浓度的芦丁标准溶液的峰面积对浓度作图，绘制标准曲线，得到回归方程，根据样品中的芦丁峰面积进行定量。

仪器与试剂

(1) **仪器**。Waters 600 高效液相色谱仪，超声仪，容量瓶，移液管，具塞锥形瓶，微量注射器（50μL，平头），滤膜（0.45μm，有机）。

(2) **色谱条件**。C_{18} 反相键合相色谱柱（250mm × 4.6mm × 5μm）；流动相为甲醇＋0.5％冰乙酸水溶液（1∶1，体积比），流速 1.0mL/min；柱温 30℃；检测波长为 254nm；进样量 20μL。

(3) **试剂**。甲醇（GR），乙酸（AR），芦丁标准品，槐米药材。

实验步骤

(1) **标准溶液的配制**

① 取芦丁标准品约 10mg，精密称定，加甲醇制成 1mg/mL 的芦丁标准品储备液。

② 分别准确吸取标准品储备液 0.50mL、1.00mL、1.50mL、2.00mL、2.50mL 于 5 个 10mL 的容量瓶中，加入流动相稀释至刻度并摇匀。

（2）样品溶液的配制

① 取槐米粉末约 0.15g，准确称定，置具塞锥形瓶中。

② 准确加入 25.00mL 甲醇，称重，置超声波清洗器中振荡提取 20min。

③ 放冷，再称重，用甲醇补足减失的重量，摇匀，过滤。

④ 精密移取滤液 5.00mL 置 100mL 容量瓶中，用流动相定容至刻度。

（3）进样分析

① 将不同浓度的芦丁标准溶液进样 20μL，分别记录其保留时间和峰面积于表 3-12 中。

② 将样品溶液进样 20μL，平行测定 3 次，记录芦丁的保留时间和峰面积于表 3-13 中。

表 3-12　不同浓度芦丁的峰面积

芦丁浓度/(mg/mL)	保留时间/min	峰面积
回归方程		
相关系数		

表 3-13　槐米中芦丁的含量计算

试样	保留时间,t_R/min	峰面积	芦丁含量/%	平均含量/%
1				
2				
3				

数据处理

（1）以芦丁的峰面积和标准品浓度作标准曲线，并得到回归方程、相关系数和线性范围。

（2）将芦丁的峰面积值代入回归方程，计算槐米中芦丁的含量。

注意事项

实验步骤中所列的色谱条件是一个参考数据。实验中，因仪器及其他因素影响，可能分离结果并不理想，可适当调整流动相的浓度及配比等，使之达到最佳分离。

思考题

（1）流动相配比改变对芦丁峰的保留时间和分离效果有何影响？

（2）测定槐米中的芦丁含量时，样品浓度为什么必须落在标准曲线范围内？

实验 35　高效液相色谱法测定头孢氨苄的含量

实验目的

（1）掌握高效液相色谱法测定头孢氨苄含量的方法。

（2）了解外标法计算药物含量的方法。

实验原理

头孢氨苄为 (6R,7R)-3-甲基-7-[(R)-2-氨基-2-苯基乙酰
氨基]-8-氧代-5-硫杂-1-氮杂双环 [4.2.0] 辛-2-烯-2-甲酸，
其分子式为 $C_{16}H_{17}N_3O_4S$，分子量 347.39，化学结构式如
图 3-8 所示。

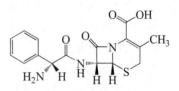

图 3-8 头孢氨苄的化学结构式

本实验采用 HPLC 的外标法对头孢氨苄胶囊中头孢氨苄
的含量进行测定，样品含头孢氨苄（$C_{16}H_{17}N_3O_4S$）应为标
示量的 90.0%～110.0%。

仪器与试剂

（1）仪器。 岛津 LC-20AD 高效液相色谱仪，紫外检测器，分析天平，移液管，容
量瓶。

（2）色谱条件。 C_{18} 反相键合相色谱柱（250mm×4.6mm × 5μm），以水-甲醇-3.86%
醋酸钠溶液-4% 醋酸溶液（742∶240∶15∶3）为流动相，流速为 0.7～0.9mL/min，检测
波长为 254nm，进样量 20μL。

（3）试剂。 头孢氨苄胶囊，头孢氨苄对照品，甲醇（AR），醋酸钠（AR），醋酸
（AR），超纯水。

实验步骤

（1）供试品溶液的制备

① 取头孢氨苄胶囊 10 粒，准确称定质量。取出胶囊，准确称定胶囊质量，并计算平均
装量。

② 将内容物混合均匀，精密称取适量（约相当于头孢氨苄 0.1g），置于 50mL 烧杯中，
加流动相适量，充分振荡，使头孢氨苄溶解。

③ 将分散液转移至 100mL 容量瓶中，再用流动相稀释至刻度，摇匀，过滤。精密量取
滤液 10mL，置 50mL 容量瓶中，用流动相稀释至刻度，摇匀，备用。

（2）对照品溶液的制备。 称量头孢氨苄对照品适量，准确配制成每 1mL 中约含 200μg
头孢氨苄的对照品溶液。

（3）进样测定

① 按岛津 LC-20AD 高效液相色谱仪的操作步骤，启动色谱仪，打开软件操作界面，设
置仪器工作参数，打开 Purge 阀排气泡，待基线平稳后进样。

② 分别精密吸取供试品溶液和对照品溶液各 20μL，注入液相色谱仪，记录色谱图，按
下式计算头孢氨苄的标示量百分数。

$$标示量百分数 = \frac{A_x \times c_R \times D \times 10^{-3} \times \overline{W}}{A_R \times W \times 每粒胶囊标示量(mg/粒)} \times 100\%$$

式中，A_x，A_R 分别为供试品溶液和对照品溶液中头孢氨苄的峰面积；D 为供试品稀释体
积；c_R 为对照品溶液的浓度（μg/mL）；W 为样品的取样量（g）；$\overline{W}$ 为平均粒重（g/粒）。

注意事项

(1) 流动相及试液使用之前，需用微孔滤膜滤过，还要进行脱气。

(2) 对照品和供试品溶液至少进样 2 次，并求其平均值。

思考题

(1) 头孢类抗生素除用 HPLC 法测定外，还有哪些方法？

(2) 试述外标法定量的原理及特点。

实验 36　高效液相色谱法测定绿茶饮料中咖啡因和茶碱的含量

实验目的

(1) 学习高效液相色谱仪的基本结构和基本操作。

(2) 了解反相液相色谱法的原理、优点和应用。

(3) 掌握高效液相色谱法进行定性、定量分析的依据。

实验原理

绿茶饮料是一种以绿茶粉末为原料的饮料，以其优异的口感，成为大众喜爱的饮品。绿茶饮料中含有茶多酚、咖啡因、茶碱、蔗糖等多种成分。咖啡因和茶碱是其中重要的生物活性物质，它能兴奋大脑皮层，使人消除疲劳。但是过量饮用会对人体造成一定程度的损害。咖啡因和茶碱都属于天然的黄嘌呤类衍生物，它们的化学名称分别为 1,3,7-三甲基黄嘌呤和 1,3-二甲基黄嘌呤。

定量测定咖啡因和茶碱的传统方法是滴定法、紫外可见分光光度法。本实验采用反相色谱法将咖啡因、茶碱和其他组分分离后，进行二极管阵列检测。在恒定的实验条件下，以色谱图上物质的保留时间 t_R 作为定性参数，以峰面积 A 作为定量参数，以不同浓度咖啡因和茶碱标准溶液的峰面积对浓度作图，绘制工作曲线。再根据未知样品中咖啡因和茶碱的峰面积，利用工作曲线法（即外标法）测定饮料中咖啡因和茶碱的含量。

仪器与试剂

(1) 仪器。岛津 LC-20AD 高效液相色谱仪，二极管阵列检测器，100mL 和 10mL 容量瓶，1.5mL 进样瓶。

(2) 色谱条件。ODS（C_{18}）柱（150mm ×4.6mm × 5μm），流动相为 70%水＋30%甲醇（体积分数），流速为 1.0μL/min，检测波长为 272nm，进样体积是 10μL。

(3) 试剂。甲醇（AR），咖啡因和茶碱标准试剂，绿茶饮料。

实验步骤

(1) 咖啡因和茶碱标准储备液的配制。准确称取 10mg 咖啡因，用配制的流动相溶解，转入 100mL 容量瓶中，稀释、定容。按照同样的方法配制茶碱标准储备液。

(2) 咖啡因和茶碱标准溶液的配制。准确移取 0.1mL 咖啡因标准储备液于 10mL 容量瓶中，用流动相定容至刻度。按照同样的方法配制茶碱标准溶液。

（3）**混合标准溶液系列的配制**。分别移取 0.10mL、0.20mL、0.30mL、0.40mL、0.50mL 的咖啡因标准储备液和等体积的茶碱标准储备液至 10mL 容量瓶中，用流动相定容至刻度。所得混合标准溶液的浓度为：1μg/mL、2μg/mL、3μg/mL、4μg/mL、5μg/mL。

（4）**标准曲线的绘制**

① 按岛津 LC-20AD 高效液相色谱仪的操作步骤，启动色谱仪，打开软件操作界面，设置仪器工作参数，打开 Purge 阀排气泡，待基线平稳后进样。

② 首先进样咖啡因标准溶液和茶碱标准溶液，确定各自的保留时间。再按照浓度从低到高的顺序进混合标准溶液，记录不同浓度的咖啡因和茶碱的峰面积。

（5）**样品测定**

① 绿茶饮料经超声波脱气 10min，0.45μm 滤膜过滤，用流动相稀释 50 倍待用。

② 进样绿茶饮料，确定饮料中的咖啡因和茶碱的保留时间和峰面积。

（6）**关机**。实验结束后，采用梯度洗脱对色谱柱进行清洗，清洗完毕后，关闭仪器和计算机。

数据处理

（1）根据标准试样色谱图中的保留时间数据，找到样品色谱图中咖啡因和茶碱的色谱峰。

（2）用标准试样的峰面积 A 对质量浓度 ρ（μg/mL）分别绘制两种分析物的工作曲线。

（3）由未知样的峰面积从工作曲线上求得其中咖啡因和茶碱的质量浓度（μg/mL）。

注意事项

（1）饮料试样必须经过脱气、过滤处理，不能直接进样。直接进样会影响色谱柱的寿命。

（2）试样和标准溶液需要冷藏保存。

思考题

（1）解释用反相色谱法测定咖啡因和茶碱的原理。

（2）高效液相色谱有哪些定性和定量分析方法？

实验 37 高效液相色谱柱的性能考察及分离度测试

实验目的

（1）掌握色谱柱理论塔板数、理论塔板高度和色谱峰拖尾因子的计算方法。

（2）掌握应用色谱图计算分离度的方法。

实验原理

评价色谱柱的性能好坏，有不同的方法和考察指标。主要指标包括色谱柱的分离度、理论塔板数、理论塔板高度、峰对称性、在不同 pH 介质中样品测定的稳定性和重现性、柱子的样品负载量等。这里主要介绍理论塔板数、理论塔板高度、峰对称性和分离度。

塔板理论把色谱柱比作精馏塔，即色谱柱是由一系列连续的、相等体积的塔板（plate）组成，每一块塔板的高度称为理论塔板高度（height of theoretical，H），它大约是被测组分在流动相和固定相之间达到一次分配平衡所需的柱长度。假设色谱柱长为 L，n 即为溶质平衡分配的次数，即理论塔板数（n）。显然，塔板数 n 愈大，板高 H 愈小，柱效能越高，这是衡量色谱柱性能的一个重要指标。本实验通过测试苯和甲苯的理论塔板数判断其柱效的高低。

色谱柱的热力学性质和柱填充得均匀与否，将影响色谱峰的对称性。色谱峰的对称性可用对称因子（f_s）或拖尾因子（T）来衡量，拖尾因子应在 $0.95\sim1.05$ 之间。

分离度 R 是从色谱峰判断相邻两组分在色谱柱中总分离效能的指标，相邻两组分的分离度应大于 1.5，才能达到完全分离。

仪器与试剂

(1) 仪器。 高效液相色谱仪（岛津 LC-20AD），C_{18} 反相键合相色谱柱（250mm × 4.6mm × 5μm），紫外检测器，50μL 微量注射器。

(2) 色谱条件。 流动相为 80% 甲醇 + 20% 水（体积分数），流速 1.0mL/min，检测波长 254nm，柱温 30℃。

(3) 试剂。 苯（AR），甲苯（AR），甲醇（GR），重蒸水。

实验步骤

(1) 配制苯、甲苯的甲醇溶液（1μg/mL）作为测试溶液。

(2) 在选定的实验条件下，样品溶液进样 20μL，记录色谱图。

数据处理

(1) 根据苯、甲苯色谱峰的 t_R（保留时间）和 $W_{1/2}$（半峰宽），按下式计算色谱柱的理论塔板数：

$$n = 5.54\left(\frac{t_R}{W_{1/2}}\right)^2$$

(2) 根据色谱峰，按下式计算各组分的拖尾因子：

$$T = \frac{W_{0.05h}}{2d_1}$$

式中，$W_{0.05h}$ 为 0.05 峰高处的峰宽；d_1 为峰极大至峰前沿之间的距离。

(3) 根据色谱图，按下式计算苯和甲苯的分离度：

$$R = \frac{2(t_{R_1} - t_{R_2})}{W_1 + W_2}$$

式中，t_{R_1} 和 t_{R_2} 分别为苯和甲苯色谱峰的保留时间；W_1 和 W_2 分别为苯和甲苯的峰宽。

注意事项

(1) 高效液相色谱仪中所用的溶剂均需纯化处理。流动相经脱气后方可使用。

（2）实验结束后，反相色谱柱需用甲醇冲洗 20～30min，以保护色谱柱。

思考题

（1）使用反相键合相色谱柱时，流动相 pH 应控制在什么范围内？

（2）用苯和甲苯表示的同一色谱柱的柱效能是否一样？

实验 38 气相色谱-质谱联用分析菜籽油的脂肪酸成分

实验目的

（1）了解气质联用仪的主要组成部件和工作原理。

（2）掌握使用 NIST（National Institute of Standards and Technology）数据库。

（3）学习识别质谱图中主要碎片离子峰，分析典型有机物结构。

（4）学习多组分混合脂肪酸的衍生化、提取方法，以及色谱分离条件的选择。

实验原理

分子质谱是试样分子在高能离子束作用下电离生成各种类型带电粒子或离子，采用电场、磁场将离子按照质荷比大小分离、依次排列成质谱。根据质谱峰的位置进行物质的定性和结构分析，根据峰的强度进行定量分析。质谱分析具有灵敏度高、定性能力强等特点；色谱法具有高效分离混合物和定量分析简便的特点。两种分析技术联用，发挥各自的专长，使分离和鉴定同时进行，具有广泛的应用领域。

气相色谱-质谱联用（gas chromatography-mass spectrometry，GC-MS）是最早实现的色谱-质谱联用技术，发展最为完善。GC-MS 色谱部分包括进样器（可以手动进样，也可以使用自动进样器）、汽化室、柱箱和载气系统。根据试样用量，可采用分流或不分流进样方式。多组分试样进入色谱柱后，由于不同组分与色谱柱固定相的相互作用不同，经过一定时间后，各组分彼此分离，先后进入质谱仪。

值得注意的是，色谱柱出口端为常压，而质谱仪在高真空度下工作，因此，如果使用的是填充柱，需要一个接口（如分子分离器）将色谱柱流出物中的载气尽可能除去。本实验使用毛细管色谱柱，可直接插入质谱仪的离子源。

GC-MS 的质谱部分包括离子源、质量分析器和检测器。本实验采用电子轰击（electron impact，EI）离子源，在 70 eV 电子轰击下，中性试样分子失去电离能低的电子，成为带电荷的分子离子，并进一步发生化学键的断裂，产生低质量的碎片离子。质量分析器的作用则是将离子源产生的离子按照 m/z 的大小分离，最后被检测器测定。质量分析器的种类很多，本实验采用四级杆质量分析器。它由四根平行的圆柱形金属电极组成，相对的电极被对角连接，构成两组电极。在两组电极之间施加数值相等而方向相反的直流电压 U_{dc} 和射频交流电压 V_{rf}，四级杆所包围的空间便产生一个双曲线形电场。当 U_{dc}/V_{rf} 一定时，只有特定 m/z 离子才能稳定振荡地通过四级杆，到达检测器，其他离子则撞到四级杆上被真空系统抽走。改变直流电压 U_{dc} 和射频交流电压 V_{rf} 可达到质量扫描的目的，获得质谱图。由于四级杆质量分析器体积小，扫描速度快，适合于色谱质谱联用仪。

仪器与试剂

(1) 仪器。安捷伦 7890B-5977B 气相色谱-质谱联用仪，Restek Rtx-5 MS 毛细管色谱柱，小型超声仪。

(2) 试剂。色谱纯正己烷，甲醇，氢氧化钠，无水硫酸钠，pH 试纸，菜籽油。

实验步骤

(1) 试样处理。称取 2mg 菜籽油置于干净的 5mL 小玻璃瓶中，加入 $100\mu L$ 甲醇（含 0.5mol/L NaOH）和 1mL 正己烷，盖上瓶盖。超声 1min 后，取出小玻璃瓶并放于冰上。等待温度降低，并且两相分离完全后，用玻璃吸管取上层正己烷，并用纯水洗至中性，加入无水硫酸钠可除去正己烷试样层中的痕量水分，最后将正己烷层置于进样瓶中，准备进样分析。

(2) 设置色谱分离条件。由于菜籽油中的脂肪酸含量较高，因此将进样方式设为分流进样，分流比为 100（不同样品，该比例可能不同，视具体情况而定）。根据试样性质，设置进样口温度、载气流速和程序升温等。

(3) 设置质谱检测条件。设置离子源温度、接口温度、检测器电压、扫描方式（全扫描或选择离子扫描）、质量范围等，特别要注意设置溶剂切除时间。

(4) 样品分析。选择手动或自动进样（进样体积一般为 $1\mu L$），获得样品谱图。采样结束后，将仪器置于待机状态。

数据处理

(1) 分析谱图。分析质谱图中主要碎片离子峰的产生机理和同位素离子峰簇特点，推导典型脂肪酸甲酯的元素组成和分子结构。

(2) NIST 数据库搜索。在 NIST 数据库中对所获得质谱图进行搜索，比较实验所得质谱图与标准谱图的差异和匹配度。

注意事项

(1) 菜籽油在碱性条件下甲酯化衍生后，溶液为碱性，需经水洗至中性并用无水硫酸钠除去残留水分，否则将造成色谱柱固定相的流失。

(2) 菜籽油中脂肪酸含量较高，若不采用分流进样方式，电流过大，灯丝可能烧断，还会损坏检测器。

思考题

(1) 为什么要设置溶剂切除时间？

(2) 饱和脂肪酸甲酯和不饱和脂肪酸甲酯的质谱图有什么显著区别？能否用质谱鉴定双键的位置？

(3) 顺式和反式脂肪酸甲酯能否用色谱-质谱联用的方式区别？

(4) 请推导饱和脂肪酸甲酯质谱图中 $m/z=74$ 碎片离子峰的产生机理。

实验 39　液相色谱-质谱联用技术对中药黄芩中黄芩苷的定性和定量分析

实验目的

（1）了解高效液相色谱-质谱联用仪的操作方法。

（2）熟悉液相色谱-质谱仪的基本工作原理。

实验原理

液相色谱-质谱联用（liquid chromatography-mass spectrometry，LC-MS）是另一种广泛应用的色谱-质谱联用技术。LC-MS 主要由液相色谱系统、连接接口、质量分析器和计算机数据处理系统组成。LC-MS 比较成熟的接口及电离技术为电喷雾电离（ESI）和大气压化学电离（APCI）。

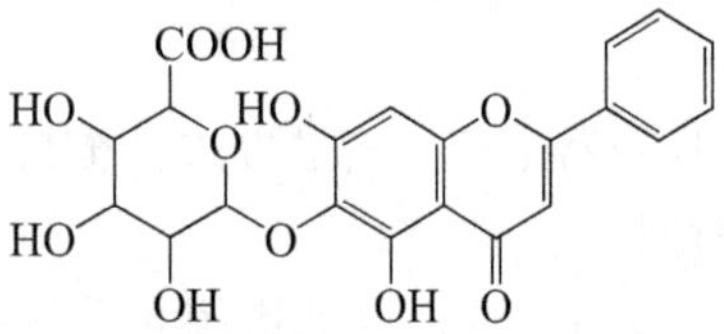

图 3-9　黄芩苷的化学结构式

黄芩的主要有效成分之一的黄芩苷为黄酮类化合物，其分子式为 $C_{21}H_{18}O_{11}$，化学结构式如图 3-9 所示。

本实验利用液相色谱-质谱技术对黄芩药材中的黄芩苷进行结构鉴定和定量分析，具有灵敏、准确、快速等特点。

仪器与试剂

（1）仪器。Agilent TOF LC/MS 联用系统，Agilent 化学工作站，分析天平，超声仪，$0.45\mu m$ 滤膜过滤器，微量注射器。

（2）试剂。黄芩药材，黄芩苷（对照品，99%），甲醇（GR），乙腈（GR），甲酸（AR），超纯水。

实验步骤

（1）仪器操作条件（参考值）

① 色谱条件。Waters Symmetry Shield RP-18（100mm×2.1mm×3.5μm），C_{18} 保护柱（10mm×2.1mm×3.5μm）；流动相：溶剂 A（水，0.1%甲酸）＋溶剂 B（乙腈，0.1%甲酸）；流速：0.4mL/min；进样量：5μL。梯度洗脱（A＋B＝100%）如表 3-14 所示。

表 3-14　梯度洗脱条件

t/min	0.00	2.00	6.00	11.00	11.25	12.25	12.50
[B]/%	30	30	60	85	99	99	30

② 质谱条件。电喷雾离子化源（ESI）；离子源喷射电压 4kV；干燥气（N_2）流速 11.5L/min；干燥气温度 350℃；雾化气（N_2）压力 2.4×10^5Pa；检测离子：一级离子 m/z 为 447.2，二级离子 m/z 为 271.1。

（2）溶液配制

① 标准储备液。准确称量黄芩苷对照品约 10mg，甲醇溶解后转移至 100mL 容量瓶中，稀释至刻度，混匀。

② 对照品溶液。准确吸取储备液适量，50%甲醇水溶液稀释，制成浓度约为 0.5μg/mL、2μg/mL、5μg/mL、10μg/mL、20μg/mL 的系列对照品溶液。

③ 试样溶液。取黄芩药材 50g 粉碎，准确称取约 0.1g，置 100mL 烧杯中，加 50%甲醇适量，超声处理 30min。放置，冷却至室温，转移至 100mL 容量瓶，以 50%甲醇稀释至刻度，摇匀。准确量取 5.00mL，置 50mL 容量瓶中，加 50%甲醇稀释至刻度，摇匀，用 0.45μm 滤膜过滤。

（3）样品分析。分别取系列对照品溶液和样品供试液，各进样 5μL，记录总离子流色谱图、黄芩苷质谱图和色谱图。试样溶液重复测定 3 次。

数据处理

（1）定性分析。黄芩苷（相对分子质量 446.35）在正离子检测方式下的 $[M+H]^+$ 峰为准分子离子峰，m/z 为 447.2；二级质谱特征离子的 m/z 为 271.1，由准分子离子 m/z 447.2 的 $[M+H]^+$ 失去一分子的葡糖醛酸产生，与对照品的质谱图和色谱保留行为对照可确认为黄芩苷。

（2）定量分析

① 以黄芩苷的峰面积和对照品浓度作线性回归，并得到回归方程和相关系数，数据填入表 3-15。

表 3-15　不同浓度黄芩苷的保留时间和峰面积

黄芩苷浓度/(mg/mL)	保留时间(t_R)/min	峰面积
回归方程		
相关系数		

② 将黄芩药材中的黄芩苷峰面积值代入回归方程，计算黄芩药材中黄芩苷的含量（表 3-16）。

表 3-16　黄芩药材中黄芩苷含量

试样	保留时间(t_R)/min	峰面积	黄芩苷含量/%	平均含量/%
1				
2				
3				

注意事项

与 GC-MS 相比，LC-MS 的系统噪声较大，应采取纯化有机溶剂、纯化样品、定期清洗系统等措施减小噪声影响。

思考题

(1) 液相色谱-质谱联用仪由哪几部分组成？其"接口"类型有哪些？

(2) LC-MS 与 GC-MS 相比有何特点？

实验 40　氨基酸的纸色谱鉴别

实验目的

(1) 掌握以纸色谱法分离鉴定混合物的原理。

(2) 掌握纸色谱的操作方法。

实验原理

色谱法按照固定相的固定方式可分为柱色谱、纸色谱和薄层色谱法。纸色谱法以滤纸作载体，纸上所吸附的水为固定相，与水不相溶的有机溶剂为流动相。主要用于极性物质的分离分析，如糖和氨基酸。在层析时将样品点在距滤纸一端约 2~3cm 的某一处，该点称为原点。随后，在密闭的容器中有机溶剂通过毛细作用穿透纸张，并穿过样品原点，从而携带样品的各种组分随流动的溶剂一起移动，其速度取决于它们在固定相和流动相中的溶解度。由于分配系数（K）不同，它们出现在滤纸的不同位置上。各物种在纸层析图谱上的位置可用比移值（retention factor，R_f）来表示，其定义为：

$$R_f = \frac{\text{原点中心至斑点中心的距离}}{\text{原点中心至溶剂前沿的距离}}$$

在相同实验条件下，R_f 值是常数，因此可用比移值进行物质的定性分析。本实验采用正丁醇-乙酸-水 4:1:1（BAW 系统）为展开剂，以上行纸色谱法分离甘氨酸（NH_2CH_2COOH）和蛋氨酸 [$CH_3SCH_2CH_2CH(NH_2)COOH$]。由于两者结构的差异，甘氨酸极性大于蛋氨酸，在滤纸上展开较慢，因而甘氨酸的 R_f 值小于蛋氨酸。展开后，氨基酸在 60℃ 下与茚三酮发生显色反应，在层析纸上形成紫红色斑点。

仪器与试剂

(1) **仪器**。带盖的层析缸，色谱纸（或滤纸），毛细管，烘箱。

(2) **试剂**。0.5% 的甘氨酸标准溶液，0.5% 的蛋氨酸标准溶液，0.5% 的甘氨酸溶液与 0.5% 的蛋氨酸溶液的等量混合液，展开剂为新鲜配制的正丁醇乙酸水（4:1:1），显色剂为 1% 的茚三酮乙醇溶液。

实验步骤

(1) **点样**。取色谱纸（15cm×5cm）一张，在距顶端 1cm 处的中间部位穿一小孔。在距色谱纸下端 1.5～2cm 处用铅笔轻轻地画一条直线。用毛细管分别点加上述标准溶液及样品混合溶液 3～4 次，斑点直径约 2mm，晾干。

(2) 展开。在干燥的层析缸中加入 35mL 展开剂，把点样后的滤纸垂直挂于层析缸内，盖上缸盖，饱和 10min，然后使滤纸底边浸入展开剂内约 0.3～0.5cm，进行展开。待溶剂前沿展开至合适的位置（约 12cm）时，用镊子取出滤纸，立即标记出溶剂前沿的位置。

(3) 显色。将滤纸晾干，喷以茚三酮显色剂，再置色谱纸于 60℃烘箱中显色 5min，或在电炉上方小心加热，即可看到紫红色的斑点。

数据处理

用铅笔绘出每个显现的斑点的轮廓，并找出各斑点的中心，计算斑点的 R_f 值，对混合样品组分进行定性分析。

注意事项

(1) 展开剂必须预先配制且充分摇匀。
(2) 每次点样用新的毛细管以防交叉污染。
(3) 茚三酮溶液对体液如汗液也能显色，在取滤纸时应用镊子夹取，以免滤纸污染。
(4) 茚三酮显色剂应临用前配制，或置冰箱中冷藏备用。
(5) 喷显色剂要均匀，适量。

思考题

(1) 影响 R_f 值的因素有哪些？
(2) 怎样才能得到层析斑点集中、溶剂前沿整齐一致的图谱？
(3) 滤纸在接触展开剂之前为什么要先置于层析缸中进行平衡？对平衡的时间及温度有何要求？
(4) 纸色谱有哪些优缺点？

实验 41　黄连药材的薄层色谱鉴别

实验目的

(1) 掌握薄层板制备方法。
(2) 掌握薄层色谱的一般操作方法。
(3) 了解薄层色谱在中药鉴别中的应用。

实验原理

薄层色谱法系将固定相涂在玻璃板或塑料板上，待样品点样后在展开剂（流动相）的作用下展开，由于被测组分和移动相和固定相具有不同的亲和力，从而产生差速迁移而得到分离。该技术广泛应用于多组分药物制剂的分离分析中。

黄连是常用中药，具有抗菌消炎的功效，含有多种生物碱，其中盐酸小檗碱为主要有效成分，在 365nm 波长紫外灯下显黄色荧光。利用薄层色谱法可将黄连中的生物碱分离，用对照药材和盐酸小檗碱对照品进行对照，可鉴别黄连药材。

仪器与试剂

（1）**仪器**。紫外分析仪，烘箱，玻璃板（5cm×10cm），研钵，双槽层析缸，小药匙，毛细管，铅笔，直尺。

（2）**试剂**。黄连药材供试品，黄连对照药材，盐酸小檗碱对照品，苯（AR），乙酸乙酯（AR），异丙醇（AR），甲醇（AR），浓氨水（CP），羧甲基纤维素钠水溶液（CMC-Na，0.7%），硅胶 G（薄层层析用）。

实验步骤

（1）**薄层板的制备**。称取 10g 薄层层析用硅胶 G 于研钵中，加入约 25mL 羧甲基纤维素钠溶液，向一个方向研磨至无气泡，使之成为均匀的混悬液。将研磨好的硅胶浆倒在洗净晾干的玻璃板上铺匀后晾干。使用前应先在 105℃烘箱中活化 30min 后置干燥器中冷却。

（2）**样品的制备**。取黄连药材供试品粉末 50mg，加盐酸-甲醇（1∶100）5mL，加热回流 15min，过滤，滤液补足溶剂至 5mL，即得供试品溶液。同法制备对照药材溶液。再取盐酸小檗碱对照品，加甲醇制成 1mL 含 0.5mg 的溶液，作为对照品溶液。

（3）**点样**。用铅笔轻轻地在离薄层板底部 1.5cm 处画一条直线为起始线，离顶端 1cm 处画线作为溶剂前沿，于起始线上分别用毛细管吸取上述三种溶液各 1μL 点样。

（4）**展开**。以苯-乙酸乙酯-异丙醇-甲醇-水（6∶3∶1.5∶1.5∶0.3）为展开剂，将薄层板置氨蒸气饱和的层析缸中进行展开，展开至溶剂前沿线时，取出薄层板。

（5）**检视**。待薄层板上的溶剂挥发后，置 365nm 紫外灯下检视，并用铅笔绘出主斑点的轮廓。

数据处理

比较三条色谱带上荧光斑点的颜色和位置。在供试品色谱带中，在与对照药材色谱相应的位置上，应显相同的黄色荧光斑点；在与对照品色谱相应的位置上，应显一个相同的荧光斑点。

注意事项

（1）薄层板的表面要均匀、平整、光滑、无麻点、无气泡，薄层厚度一般 0.2～0.3mm 为宜。铺好的板一定要晾干后才能活化，以防开裂。

（2）薄层板在烘箱中活化后应立即放入干燥器中冷却。

（3）点样量应适当，太少则斑点不明显，太多则出现拖尾，点样要轻不能刺破薄层板，点样直径一般不大于 2～3mm。

（4）点样间距可视斑点扩散情况以相邻斑点互不干扰为宜，一般不少于 8mm。

（5）层析缸必须密封，否则溶剂易挥发，从而改变展开剂的比例，影响分离效果。

思考题

（1）薄层显色的方法有哪些？

（2）物质发生荧光的条件是什么？

(3) 为什么层析缸中要加入氨蒸气饱和？

实验 42　离子色谱法测定自来水中的阴离子

实验目的

(1) 了解离子色谱分析的基本原理和操作方法。
(2) 掌握离子色谱法的定性和定量的分析方法。

实验原理

离子色谱法（ion chromatography，IC）是在离子交换色谱基础上派生出来的一种液相色谱方法，它是将色谱法的高效分离技术和离子的自动检测技术相结合的一种分析方法。离子色谱法以离子交换树脂为固定相，电解质溶液为流动相，通常采用电导检测器来进行检测。离子色谱仪有单柱型和双柱型，一般均由四个部分组成，即输送系统、分离系统、检测系统和数据处理系统。

本实验以阴离子交换树脂为固定相，以 $NaHCO_3$-Na_2CO_3 混合液为洗脱液，分析水中 Br^-，NO_3^- 和 SO_4^{2-} 三种阴离子。当含待测阴离子的试液进入分离柱后，在分离柱上发生如下交换过程：

$$RHCO_3 + MX \Longleftrightarrow RX + MHCO_3$$

式中，R 代表离子交换树脂。

由于洗脱液不断流过分离柱，使交换在阴离子交换树脂上的各种阴离子又被洗脱，而发生洗脱过程。由于不同的阴离子与离子交换树脂的亲和力的不同，交换和洗脱过程有所不同，亲和力小的离子先流出分离柱，而亲和力大的离子后流出分离柱，因而各种不同的离子得到分离。

当待测阴离子从柱中被洗脱而进入电导池时，要求电导检测器能随时检测出洗脱液中电导的改变。但因洗脱液中 HCO_3^-、CO_3^{2-} 的浓度比试样阴离子的浓度大得多，因此与洗脱液本身的电导值相比，试液离子的电导贡献显得微不足道，因而电导检测器难以检测出由于试液离子浓度变化所导致的电导变化。对于具有抑制柱的离子色谱，来自再生液中 H^+ 通过阳离子交换膜进入淋洗液，与淋洗液的 CO_3^{2-}、HCO_3^- 和 X^- 结合形成弱电离的 H_2CO_3 和强电离 HX。为了保持淋洗液和再生液的电中性，化学计量的 Na^+ 向相反方向移动，即从淋洗液通道到再生液，最后被带入废液，结果使洗脱液中 $NaHCO_3$ 和 Na_2CO_3 转化成 H_2CO_3，大大降低了本底电导。而试样中 MX 转化为相应的酸 HX，由于 H^+ 的离子淌度是金属离子 M^+ 的 7 倍，因而使得试液中离子电导的测定得以实现。

仪器与试剂

(1) 仪器。Metrohm 861 型离子色谱仪，IC Net 2.3 色谱工作站，Metrosep A supp 4 阴离子交换柱（250mm×4.0mm i.d.），Metrohm MSM Ⅱ 抑制器＋853 型 CO_2 抑制器，电导检测器。

(2) 试剂

①称取 19.10g Na_2CO_3（分析纯）和 14.30g $NaHCO_3$（分析纯）（均已在 105℃烘箱中

烘 2h 并冷却至室温），溶于高纯水中，转入 1000mL 容量瓶中，加水至刻度，摇匀。然后将此淋洗储备溶液存于聚乙烯瓶中，在冰箱中保存。此淋洗储备溶液为 0.18mol/L Na_2CO_3 + 0.17mol/L $NaHCO_3$。②分别配制浓度为 1000mg/L 的 Br^-、1000mg/L 的 NO_3^-、1000mg/L 的 SO_4^{2-} 的阴离子标准溶液，测定时稀释为标准使用溶液。混合标准使用溶液为含有 20mg/L Br^-，20mg/L NO_3^- 和 200mg/L SO_4^{2-} 的水溶液，测定时配制。

实验步骤

（1）阴离子淋洗液的制备。移取 0.18mol/L Na_2CO_3 + 0.17mol/L $NaHCO_3$ 阴离子淋洗储备溶液 10.00mL，用高纯水稀释至 1000mL，摇匀。此淋洗液为 1.8mmol/L Na_2CO_3 + 1.7mmol/L $NaHCO_3$。

（2）开机。依次打开离子色谱的电源开关，IC Net 2.3 色谱工作站，启动泵，调节流动相流速为 1mL/min，使系统平衡 30min，等待仪器稳定，色谱流出曲线的基线平直。

（3）单个标准溶液分析。将仪器调至进样状态，启动 Fill 键，用注射器吸取 1mL 各阴离子标准溶液进样。再启动 Inject 键，开始进行色谱分析，待峰全部出完后，记录各个阴离子的保留时间。

（4）混合标准溶液分析。取混合阴离子标准溶液，按照步骤（3）直接进样，从步骤（3）的几个阴离子的保留时间可确认混合标准溶液的中峰所对应的阴离子。

（5）工作曲线的绘制。分别取阴离子混合标准液 0.50mL、1.00mL、2.00mL、3.00mL、4.00mL 于 5 个 10mL 容量瓶中，用高纯水稀释至刻度，摇匀。每种溶液分别进样 2 次，记录色谱图。以离子浓度对峰面积作图，绘制各离子的工作曲线。

（6）自来水样品测定。取实验室自来水样，经 0.45μm 微孔滤膜过滤后在同样的实验条件下重复进样 2 次，记录色谱图。由色谱峰的保留时间定性，由色谱峰面积计算自来水中各离子的含量。

思考题

（1）简述离子色谱的分离机理。

（2）为什么需要在电导检测器前加入抑制器？

实验 43　毛细管区带电泳法分离氧氟沙星对映异构体

实验目的

（1）熟悉毛细管区带电泳法的基本原理和方法。

（2）了解毛细管区带电泳法在拆分手性药物中的应用。

实验原理

毛细管电泳（capillary electrophoresis，CE）是一类以高压直流电场为驱动力，以毛细管为分离通道，依据试样中各组分淌度和分配行为的差异而实现分离分析的新型液相分离分析技术。由于其高效、快速、微量等优点，近年来在手性分析领域引起广泛关注。毛细管区带电泳（capillary zone electrophoresis，CZE）是毛细管电泳中最基本也是应用最广泛的一种操作模式。

氧氟沙星（ofloxacin，OFLX）是广谱的喹诺酮抗菌药，临床使用的为外消旋体的 OFLX，以及光学纯体的左旋氧氟沙星（levofloxacin，S-OFLX），商品名为 Cravit。研究表明，左旋氧氟沙星（图 3-10）的抗菌活性是右旋的 8～128 倍，是消旋体的 2 倍。其作用机理是通过抑制细菌 DNA 旋转酶（细菌拓扑异构酶Ⅱ）的活性，阻碍细菌 DNA 的复制而达到抗菌作用。左旋氧氟沙星的化学结构如下。

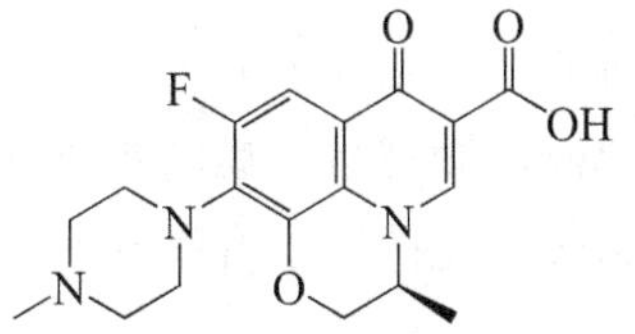

图 3-10　左旋氧氟沙星的化学结构式

本实验采用二甲基-β-环糊精（DM-β-CD）为手性选择剂，利用毛细管区带电泳法实现氧氟沙星的手性拆分。

仪器与试剂

(1) 仪器。 毛细管电泳仪，pH 计，熔融石英毛细管，分析天平，$0.45\mu m$ 微孔滤膜。

(2) 电泳条件（参考值）。 环糊精浓度为 40mmol/L，电泳缓冲液是 70mmol/L KH_2PO_4（H_3PO_4 调节 pH 2.5），分离电压为 20kV，检测波长为 280nm，柱温 25℃。

(3) 试剂。 氧氟沙星药品，左旋氧氟沙星药品，DM-β-CD（AR），KH_2PO_4（AR），H_3PO_4（AR），0.1mol/L NaOH 溶液。

实验步骤

(1) 支持电解质溶液的制备。 配制含有 40mmol/L 二甲基-β-环糊精的 KH_2PO_4（70mmol/L）溶液，用 H_3PO_4 调节其 pH 为 2.5。$0.45\mu m$ 微孔滤膜过滤后，超声波脱气备用。

(2) 试样溶液的制备。 分别精确称取氧氟沙星和左旋氧氟沙星适量，加入二次去离子水溶解，配制成 1.4mg/mL 的溶液，$0.45\mu m$ 微孔滤膜过滤，并用超声波脱气备用。

(3) 进样分析。 毛细管在每次分析之前依次用 0.1mol/L NaOH 溶液、二次去离子水冲洗 10min，然后用支持电解质缓冲溶液冲洗 5min，加电压平衡 10min。氧氟沙星进样量：1.4kPa，进样 1s，按前述条件对氧氟沙星样品进行分析。再加入左旋氧氟沙星，进一步确认左旋氧氟沙星和右旋氧氟沙星峰的位置。按前述条件分析左旋氧氟沙星样品，进样量：13.8kPa，进样 1s。

数据处理

(1) 记录氧氟沙星和左旋氧氟沙星样品的手性分离电泳图谱。

(2) 按面积归一化法分别计算氧氟沙星和左旋氧氟沙星样品中左旋体和右旋体所占的比例。

注意事项

(1) 在毛细管电泳中，环糊精种类、浓度、电解质浓度、柱温和电压等是影响分离的重要因素，实验中应优化这些条件改善分离度。

(2) 二甲基-β-环糊精在 30～40mmol/L 可达到最好的分离效果，应控制其浓度在 30～40mmol/L 范围内。

思考题

(1) 毛细管电泳法的分离机制和特点是什么？

(2) 如何保证毛细管电泳迁移时间和分析结果的重复性？

(3) 在毛细管电泳中，影响谱带展宽的因素有哪些？

Chapter 1
Fundamentals of Instrumental Analysis Experiments

I   Laboratory Safety

(1) In any laboratory safety is paramount. Take note of the location of safety showers, eye wash stations and fire extinguishers when entering the lab. Safety glasses and lab coats are required and must be worn at all times in the laboratory, no exceptions.

(2) All chemical waste should be disposed of in a properly labeled waste container. Be sure you have the correct labeling when you throw away the waste. Accidents can happen if wastes are improperly mixed.

II   Basic Requirements for Instrumental Analysis Experiment

Instrumental analysis experiment is an important content of instrumental analysis course, which is generally held simultaneously with the lecture. It is a teaching practice in which students use analytical instruments to obtain the chemical composition, structure and content of unknown substances under the guidance of teachers. Through the study of this course, students can deepen their understanding of the basic principles of related instrumental analysis methods and master the basic operations and skills of instrumental analysis experiments. They can learn to use analytical instruments correctly, choose the experimental conditions reasonably according to the the analysis requirements of real samples. Through learning, students can also correctly process data and express experimental results. This course can also cultivate their rigorous scientific attitude of seeking truth from facts, the courage to innovate in technology, and the ability to work independently. To achieve the teaching objective above, the following basic requirements are put forward:

(1) **Be well prepared.** Generally, the instruments used for instrumental analysis are expensive, so it is impossible to purchase multiple sets of similar instruments in the same laboratory. Thus the teaching of instrumental analysis experiments is usually organized in a large group. In many cases, the content of the lecture course has not yet been learned, and the experiment has already begun. Therefore, students must do the preview carefully before

the experiment. Read carefully the instrument analysis experiment textbook and lecture textbook, clarify the purpose of the experiment, and thoroughly understand the relevant principles of the experiment. Watch carefully the related online experimental videos, understand the operating procedures of the instrument, and write a pre-study report.

(2) Use the instrument correctly. Run the instrument under the guidance of the teacher. Don't turn on/off the instrument without the permission of the teachers, and don't rotate the instrument knob or change the working parameters of the instrument at will. If the instrument is found to be working abnormally during the experiment, report it to the teacher in time. Understand the performance of the instrument in detail to prevent damage to the instrument or safety accidents. Keep clean and quiet in the laboratory at all times.

(3) Develop good experimental habits. During the experiment, carefully observe the experimental phenomenon and record the experimental conditions. Truthfully analyze the raw data of the test. After the experiment, restore the instrument used. Wash used utensils timely and always keep the laboratory neat and orderly.

(4) Write the experiment report carefully. The lab report should be concise and the diagrams and tables should be clear. The lab report generally includes the following items: experiment name, the date, objectives, principles, instrument and reagents, content and procedures, experimental data or figures, data analysis and processing, and questions and discussions.

Ⅲ Data Processing and Results Expression

(Ⅰ) Basic criteria to evaluate analytical methods and results

Quantitative analysis is one of the main tasks of instrumental analysis. The evaluation of quantitative analysis methods and their analysis results requires certain performance parameters and characteristics. Generally, quantitative analysis methods have the following common performance parameters and characteristics.

(1) Standard curves and the linear range. A commonly used method for quantitative analysis is the standard curve method. The standard curve, also known as the calibration curve, refers to the relationship between the concentration (or content) of the measured substance and the response signal of the instrument.

The range of the concentration (or content) of the measured substance corresponding to the linear portion of the standard curve is known as the linear range of the method. In general, the wider the linear range, the better the analysis method.

The standard curve is obtained based on the concentration (or content) of the standard series and the corresponding response signal. Due to the random error, there is an average deviation $\bar{d}$ between a single measurement value (x or y) and the average value of n measurements ($\bar{x}$ or $\bar{y}$). According to the least-squares principle, the method of studying the relationship between dependent and independent variables is called regression analysis. If there is only one independent variable, it is called unary linear regression analysis. Let the concentrations of standard series be x_1, x_2, $\cdots$, x_i, $\cdots$, x_n respectively, the measured values of the response signals are y_1, y_2, $\cdots$, y_i, $\cdots$, y_n. If the sum of the square deviation $(\sum d^2)$ of each point to a certain straight line is the smallest or zero, then such a straight line is the best-fit unary regression straight line. According to the principle, set a unary linear equation

$$y = bx + a \tag{1-1}$$

Solve formula (1-1) and then get

$$b = \frac{\sum_{i=1}^{n}(x_i - \overline{x})(y_i - \overline{y})}{\sum_{i=1}^{n}(x_i - \overline{x})^2} \tag{1-2}$$

$$a = \overline{y} - b\overline{x} \tag{1-3}$$

Where $\overline{x} = \sum_{i=1}^{n} x_i / n$, $\overline{y} = \sum_{i=1}^{n} y_i / n$.

From formulas (1-1), it can be seen that when $x = 0$, $y = a$; when $x = \overline{x}$, $y = \overline{y}$. Draw a straight line within the concentration range of x through two points $(0, a)$ and $(\overline{x}, \overline{y})$, and this line is the optimal standard curve determined by the given set of data (x_i, y_i).

It is of practical significance to judge whether the linear relationship of this standard curve is built or not, which can be tested by the correlation coefficient r. r is a statistical parameter that characterizes the degree of correlation between the variables.

$$r = \pm \frac{\sum_{i=1}^{n}(x_i - \overline{x})(y_i - \overline{y})}{\left[\sum_{i=1}^{n}(x_i - \overline{x})^2 \sum_{i=1}^{n}(y_i - \overline{y})^2\right]^{1/2}} \tag{1-4}$$

The value of r is between $+1$ to -1. When $|r| = 1$, there is a strictly linear relationship between y and x and all y values lie on the same straight line. When $r = 0$, there is no linear relationship between y and x. When $0 < |r| < 1$, there is a certain linear relationship between y and x. That is, the closer the $|r|$ is to 1, the better the correlation between y and x.

(2) Detection limit and sensitivity. The detection limit of the analytical method refers to the minimum concentration or the minimum mass of the measured substance that can be detected at an appropriate confidence level (usually a confidence level of 99.7%). The limit of detection (LOD) is calculated by the minimum detection signal value and blank noise as shown in Eq. (1-5). The units of minimum detection concentration and minimum detection mass are expressed in $\mu g/mL$, ng/mL and μg, ng, and pg, respectively.

$$\text{LOD} = \frac{X_L - \overline{X}_b}{S} = \frac{3s_b}{S} \tag{1-5}$$

Where X_L is the minimum signal value to be detected, $\overline{X}_b$ is the average value of blank signal obtained by multiple measurements, and s_b is the standard deviation of blank signals. S is the slope of the standard curve in the low concentration area, which indicates the sensitivity of the analytical method, that is, the degree of analytical signal change when the concentration of the measured component changes by one unit.

In instrumental analysis, the sensitivity of the analytical method directly depends on the sensitivity of the detector and the magnification of the instrument. As the sensitivity increases, so does the noise. But the ratio of signal to noise (S/N) and the detection ability of the analytical method do not necessarily improve. If only the sensitivity is given without any instrumental conditions to obtain such sensitivity, the detection capability between the various analytical methods is not comparable. Since sensitivity does not take into account the influence of measurement noise, it is now recommended to use detection limit instead of sensitivity to represent the detection ability of analytical methods.

(3) **Accuracy.** Accuracy refers to the degree to which the measured value x agrees with the true value or standard value μ under certain experimental conditions. It represents the system error, expressed in terms of error or relative error E_r. The smaller the error or relative error, the higher the accuracy [Eq. (1-6)].

$$E_r = \frac{x - \mu}{\mu} \times 100\% \tag{1-6}$$

In practice, standard substances or standard methods are usually used for controlled trials. When there is no standard substance or standard method, the recovery test of pure substance added to the measured component is commonly used to estimate and determine the accuracy, that is, to measure its recovery. It is important to note that the recovery is used to estimate the accuracy of the measurements only when the systematic error varies with concentration.

When the error is small, the average value $\overline{x}$ of multiple parallel measurements is close to the true value μ, so $\overline{x}$ is often used as the estimated value of μ in practice.

(4) **Precision.** Precision refers to the degree to which the measured values obtained from multiple parallel measurements of the same sample with the same method agree with each other. It represents the random error while measuring, which is also known as repeatability. Precision is usually represented by standard deviation s or relative standard deviation s_r.

$$s = \left[\frac{1}{n-1} \Sigma (x_i - \overline{x})^2 \right]^{1/2} \tag{1-7}$$

$$s_r = \frac{s}{\overline{x}} \times 100\% \tag{1-8}$$

Where x_i is the measured value of each time, $\overline{x}$ is the average value of n measurements, and n is the number of repeated measurements.

(5) **Selectivity.** Selectivity is the ability of an analytical method to avoid interference from other coexisting components when determining a component in a sample. Selectivity is usually expressed as the ratio of the allowable amount (concentration or mass) of the coexisting component to the amount (concentration or mass) of the measured component at the specified measurement accuracy. The larger ratio indicates that the method is resistant at the specified accuracy, namely, the better selectivity.

(6) **Response time and analysis efficiency.** The response time of an instrument refers to the time required for detection signal to reach a certain percentage of the total change under

the stimulation of the sample with the excitation signal. For example, the response time of an ion-selective electrode is the time required for the potential change of the ion-selective electrode and the reference electrode to be stable ($\pm 1mV$) from contacting the test solution. In general, the shorter the response time, the better for the instrument.

Analytical efficiency (speed) is the number of samples that can be measured per unit of time. In general, the more efficient the analysis, the better the analytical method.

(II) Data analysis and result expression

(1) Reading and expression of measured data. In instrumental analysis, the original signal related to chemical information is generally converted into an electrical signal, which is amplified and then displayed directly in digits. To ensure the accuracy of measurement, the displayed value must be read correctly.

(2) Data analysis and results expression. Expressions of analysis data and results mainly include tabulation method and mathematical equation method. Basically, they should be accurate, clear and convenient for use.

① Tabulation method. Tabulation method refers to representing data in a table, which is intuitional, concise, and often used to record experimental data. The tabulation should be marked with the name of the table. The column is generally the test number or dependent variable, while the row is the independent variable. The name and dimension should be written at the beginning of the row or column. The name should be represented by symbols as far as possible, and the unit should be separated by a slash. For example, to express the temperature T, "T/K" should be written at the beginning. The recorded data should meet the requirements of significant figures. The writing should be neat and uniform, and the decimal points should be aligned up and down to facilitate the comparison and analysis of the data. If certain data in a table needs a special explanation, a mark can be made in the upper right corner of the data, such as, and the explanation is added at the bottom of the tabulation.

② Mathematical equation method. In most cases of instrumental analysis, it is a relative measurement, which needs a standard curve for quantitative analysis. Because of the inevitable measuring error, all data points rarely lie on the same straight line. Especially when the measuring error is large, it is difficult to obtain a reasonable standard curve with a simple method. In this case, it is more appropriate to describe the relationship between independent and dependent variables by the mathematical equation method. For example, the best-fit unary regression line and the unary linear equation can be obtained by plotting software such as EXCEL or ORIGIN.

Chapter 2
The Structure and Operation of Instruments

| The Components and Fundamentals of Spectrometers

(I) 722 spectrophotometer

(1) Performance and structure. 722 spectrophotometer is a single-beam instrument used in the visible spectrum area. Its working wavelength range is 360-800nm. It uses a tungsten filament incandescent lamp as the light source, a prism as a monochromator, a self-aligning optical path, a GD-7 vacuum phototube as the photoelectric converter, a field effect tube as the amplifier, and a microammeter to display the microcurrent. The image of 722 spectrophotometer is shown in Figure 2-1.

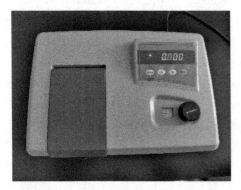

Figure 2-1　Photograph of 722 spectrophotometer（color picture）

(2) Operation steps

① Turn on the power supply, warm up the instrument, and select the detection MODE to "T" mode.

② Open the sample chamber lid（the light door closes automatically）and press the "0％T" key to make the number display "0.000".

③ Rotate the wavelength handwheel to adjust the wavelength required for the test to the tick mark.

④ Load the blank solution and the sample solution into the cuvette. Place them into the cuvette holder in the sample chamber, respectively. Generally, the blank cuvette is placed in the first holder.

⑤ Close the sample chamber lid, pull the cuvette containing the reference solution into the optical path, and press the "100％ T" key to make the number appear as 100.0.

⑥ Pull the measured solution into the optical path, and directly read the transmittance (T) value of the measured solution on the digital table.

⑦ For the measurement of absorbance A, adjust the instrument "0. 000" and "100. 0" according to steps ② and ⑤. Select the measurement mode to A so that the number is displayed as "0. 000". Pull the liquid to be measured into the optical path, and the value displayed on the screen is the absorbance A of the sample.

(3) Notes

① To achieve thermal equilibrium inside the instrument, the warm-up time should not be less than 30 minutes.

② If the continuous measurement time is too long, the phototube will fatigue and cause the reading to drift. Therefore, the sample chamber lid should be opened after each reading (the shutter closes automatically).

(Ⅱ) T9 ultraviolet-visible spectrophotometer

(1) Performance and structure. T9 double-beam ultraviolet-visible (UV-Vis) spectrophotometer has a special grating and ultra-low noise signal detection system. The detection range is -6-6, which fully meets the test requirements of high absorbance samples. It adopts a discrete three-slit combination with a continuous variable slit design, which can automatically scan the spectral bandwidth in the range of 0. 1-5. 0nm. It can identify the spectral bandwidth when the resonance absorption of the sample molecules is at its strongest and finally set the correct experimental conditions. The image of T9 UV-Vis spectrophotometer is shown in Figure 2-2.

Figure 2-2 Photograph of T9 UV-Vis spectrophotometer (color picture)

(2) Operation steps

① Machine startup. Switch on the instrument, and then turn on the computer. Double-click the software UVWIN T9CS and wait for the instrument's initial self-test.

② Spectrum scanning

a. Click "Spectrum scan" on the operation interface, set the scanning wavelength range, and click "OK".

b. Fill the pair of cuvettes with blank solution, clean the clear sides of the cuvettes, and place them into the cuvette holders.

c. Close the lid, and click the "Baseline" button to do the auto-zeroing.

d. After the baseline correction is completed, replace the sample cuvette with the solution to be measured and remain the reference cuvette unchanged. Click the "start" button to scan the spectrum.

e. Read the maximum absorption wavelength.

f. Export data and save it.

③ Standard curve plotting and unknown sample determination

a. Click "Quantitative determination" and set proper parameters, including quantitative determination wavelength and the number of repetitions.

b. Click the "Star" button and input the concentration value of the sample, and the instrument automatically reads the absorbance value.

c. Change the standard solutions of different concentrations in turn, click "start" and read the absorbance value one by one.

d. The system automatically displays the standard curve.

e. Put in the sample, and read the absorbance value and the corresponding concentration value.

④ Switching off. Save the data, take out the cuvettes, exit the software, and power off the machine.

(3) Notes

① Do not operate the instrument during the self-test.

② During baseline correction, ensure that there are no obstacles on the sample and reference beam, and there are no samples in the sample chamber.

③ When changing the sample, the cuvette needs to be rinsed with the solution to be measured 2-3 times.

(Ⅲ) LS 45 fluorescence spectrophotometer

(1) Performance and structure. LS 45 fluorescence spectrophotometer can measure fluorescence, phosphorescence, bioluminescence or chemiluminescence. The excitation slit is

Figure 2-3　Photograph of LS 45 fluorescence spectrophotometer (color picture)

2. 5-15nm and the emission slit is 2. 5-20nm. The machine uses a pulsed xenon lamp as the light source. It has a long service life and a simple power supply. It does not need a long-time preheating and can greatly reduce photolysis. Phosphorescence detection can be achieved by software control without any accessories. The picture of LS 45 fluorescence spectrophotometer is shown in Figure 2-3.

(2) Operation steps

① Solutions preparation. Prepare the series of standard solutions and the sample solutions as required.

② Machine startup. Switch on the instrument, turn on the computer and open the specific software.

③ Excitation and emission spectra scanning

a. Put the reference solution into the sample holder, click "Determine", and click "Auto Zero".

b. Put one of the standard solutions into the sample holder. Set the emission wavelength and scan the excitation spectrum of the standard solution within a specific wavelength range.

c. Set the excitation wavelength and scan the emission spectrum of the standard solution within a specific wavelength range.

④ Sample determination

a. Set the proper parameters.

b. Measure the fluorescence intensity of the series of standard solutions and the tested samples in turn.

c. Save the data.

d. Plot the standard curve of fluorescence intensity versus concentration and calculate the concentration of the unknown sample using the linear equation.

⑤ Switching off. Exit the test system, and power off the instrument and computer after cooling the machine for 30 minutes.

(Ⅳ) FTIR Nicolet iS50 spectrometer

(1) Performance and structure. The Nicolet iS50 Fourier transform infrared (FTIR) spectrometer (Figure 2-4) has an all-in-one material analysis workstation with dedicated accessories and integrated software. The high-precision mirror positioning technology of its Vectra interferometer breaks through the performance limits of traditional pneumatic or mechanical interferometers. It is not affected by vibration and noise and provides the best spectral detection technology for applications such as fast continuous scanning, step scanning and dual-channel detection.

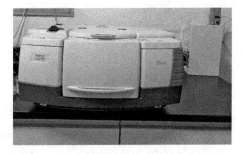

Figure 2-4　Photograph of FTIR Nicolet iS50 spectrophotometer (color picture)

(2) Operation steps

① Turn the spectrometer on.

② Turn on the computer and double-click the OMNIC icon when the optical station self-diagnosis is completed.

③ Click Experiment Setup and click Collect to collect data.

a. Select the Bench tab to check the strength of the interference signal, which should be about 6 V (Max).

b. Click Collect, select the condition parameter (usually choose the first or third background collection mode), and then click OK.

④ Acquire spectra

a. When the background collection mode is the third one, place the pressed blank KBr pellet into the sample holder and then into the sample hatch. Click Background Collection.

b. Place the KBr pellet containing the sample into the sample holder and then into the sample hatch. Click Sample Collection. In the dialog box that appears, enter the file name, and click OK, OK. When data collection is finished, select YES in the window that appears in the upper left corner, and the spectrum is displayed in the Spectra view.

c. When the background collection mode is the first type, collect the background and sample data according to the screen tips.

⑤ Spectral processing

a. Go to the "A" mode and click Aut Bsln for automatic baseline correction and return to the "T" mode.

b. Click Find pks, and then click Replace to replace the original spectrum.

c. Click Print to print the spectrum.

d. Click File—Save As, select Path, enter the filename, and then click Save.

e. Click Clear to clear not needed spectra.

(3) Notes

① The interferometer sealing chamber and the sample chamber should be placed with sufficient and effective desiccant (the color of silica gel indicator shall not turn red).

② All changes to the plug must be made when the power is cut off.

③ Ensure that the entire system must be reliably grounded. If the power supply fluctuates greatly, please add a purifying and regulated power supply. A manual reset is required after a power failure.

④ Keep the working environment temperature between 15-25℃ and the relative humidity < 60%.

(Ⅴ) GGX-810 AAS

(1) Performance and structure. GGX-810 atomic absorption spectrophotometer (AAS) adopts an 8-lamp vertical turret system, which can rotate automatically. The machine can

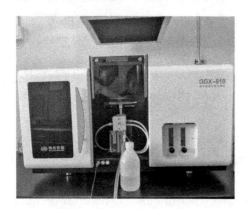

carry out analysis immediately after switching the element lamp. It can also realize automatic wavelength positioning with an accuracy of less than ±0.1nm. The floating optical platform design of the machine ensures the high stability of the optical system and reduces the influence of vibration and mechanical stress changes. The image of GGX-810 AAS is shown in Figure 2-5.

Figure 2-5   Photograph of GGX-810 AAS (color picture)

(2) Operation steps

① Machine startup

a. Turn on the computer, open the air compressor, and switch on the instrument.

b. Open the instrument software "AAS. exe".

② Determination of the absorbance of standard solution and unknown sample

a. Create a new file name, select the element to be measured, set the peak, auto adjust lamp position and zero.

b. Open the method library and open the "Element to be measured".

c. Turn on the acetylene cylinder and start to "fire".

d. Click "Analysis test", and insert the quality control sample. Change the number of measurements to "3".

e. Spray deionized water into the instrument, click "Standard blank", click "Measure", and the instrument will do the auto-zeroing.

f. Spray in standard solutions of different concentrations in turn (from dilute to concentrated solution), measure and record the absorbance value.

g. Spray in the deionized water to clean the air circuit.

h. Spray in the unknown sample and record its absorbance value.

i. After the test, spray in the deionized water to clean the air circuit.

③ Switching off

a. Close the hollow cathode lamp and the main valve of the acetylene cylinder. After the acetylene in the pipe is burned out, close the gas valve of acetylene and the air compressor.

b. Exit the software and power off the machine.

(3) Notes

① The instrument should be placed in a dry room and on a firm and stable workbench. The indoor lighting should not be too strong. In hot weather, the fan cannot blow directly to the instrument, to prevent the instability of the bulb filament from shining.

② Before using the instrument, the user should first understand the structure and working fundamentals of the instrument, and the functions of each control knob. Before powering the machine, the safety performance of the instrument should be checked.

③ When the instrument is not powered on, the ammeter pointer must be at the "0" mark. Otherwise, it can be adjusted with the correction screw on the ammeter.

(Ⅵ) 5100 ICP-OES

(1) Performance and structure. 5100 inductively coupled plasma optical emission spectrometer (ICP-OES) uses the intelligent spectrum combination technology of a synchronous vertical dual view (SVDV) system. It can obtain the horizontal and vertical observation results of plasma in one reading, reducing the number of measurements and argon consumption. The VistaChip Ⅱ detector is a CCD detector that has a high-speed continuous wavelength coverage and consumes zero gas. It delivers fast warm-up, high throughput, high sensitivity, and the largest dynamic range. Cold cone interface (CCI) minimizes self-absorption and recombination interference by removing the tail flame of cold plasma from the axial optical path. A solid-state RF system delivers a reliable, robust and maintenance-free plasma from long-term analytical stability. The picture of 5100 ICP-OES is shown in Figure 2-6.

(2) Operation steps

① Machine startup

a. Confirm that there is enough argon for continuous operation.

b. Confirm that the waste liquid drum has enough space to collect waste liquid.

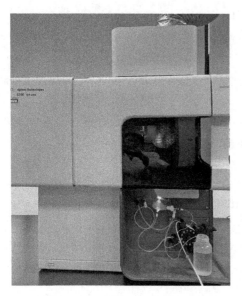

Figure 2-6   Photograph of 5100 ICP-OES (color picture)

c. Turn on the regulated power switch and ensure the power supply is stable.

d. Open the argon cylinder and adjust the partial pressure to 0.60-0.65MPa. Ensure that the instrument is gas purged for more than 1h.

e. Turn on the computer.

f. Turn on the machine power to start warming up.

g. After the instrument self-test is completed, double-click the "iTEVA" icon to start the iTEVA software, enter the main interface of the operation software, and the instrument starts initialization.

② Analysis method editing

a. Select the measuring elements and specific analysis lines.

b. Set proper parameters.

c. Set parameters of the working curve.

③ ICP torch igniting

a. Reconfirm the argon reserve and pressure and ensure that the gas purge time is longer than 1h, to prevent the CID detector from frosting and damage.

b. The light chamber temperature is stable at $38 \pm 0.29℃$. CID temperature is less than $-40℃$.

c. Check and confirm whether the sample injection system (torch tube, atomization chamber, atomizer, pump tube, etc.) is installed correctly.

d. Clamp the peristaltic pump and put the sample tube into deionized water.

e. Turn on ventilation.

f. Turn on circulating cooling water.

g. Open the "plasma status" dialog box in iTEVA software to check whether the interlocking protection is normal. If there is a red light warning, it is necessary to check accordingly. If everything is normal, click "plasma on" to ignite the ICP torch.

h. After the plasma is stable for 15min, the sample can be measured.

④ Standard curve plotting and sample detection.

⑤ ICP torch extinguishing

a. Put the sample injection tube into deionized water and flush the sample injection system for 10min after analysis.

b. In the "plasma status" dialog box, click "plasma off" to close the ICP torch.

c. After extinguishing the ICP torch for 5-10min, close the circulating water, loosen the pump clamp and pump pipe, and take out the sample injection tube from the deionized water.

d. Shut off the ventilation.

e. When the CID temperature rises to above 20℃, turn off the argon after purging for 20min.

II The Components and Fundamentals of Electrochemical Instruments

(I) pHS-3C pH meter

(1) Performance and structure. The pHS-3C pH meter adopts the AD515J operational amplifier with high input impedance as the impedance conversion, while the direct-current millivolt potential and pH value are displayed by a three-digit semi-digital voltmeter. The main part of the instrument adopts integrated circuits, with a simple design and high reliability. The picture of pHS-3C pH meter is shown in Figure 2-7.

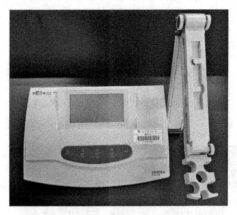

Figure 2-7　Photograph of the pHS-3C
pH meter (color picture)

(2) pH measurement in solution

① Electrode installation. Insert the pH glass composite electrode into the potentiometer jack and then turn on the instrument for preheating.

② Calibration

a. Remove the protective cap of the pH composite electrode, wash the electrode with deionized water, and then clean the surface of the electrode with filter paper.

b. Immerse the electrode in a standard solution with a pH of 6.86, turn on the stirring switch, measure the temperature of the solution with a thermometer, and then set this temperature on the instrument.

c. After the displayed value is stable, press the "Position" button. When the display shows "Std Yes", press the "Confirm" button. After that, press the "Confirm" button again, and the instrument will recognize the pH value of the standard solution at the current temperature.

d. Take out the electrode and clean it with filter paper, then immerse it in the second standard solution (according to the acid-base condition of the test solution), press the "Slope" button and "Confirm" button successively, the instrument will recognize the pH value of the current solution at the current temperature.

e. The calibrated instrument can measure the pH value of the unknown solution. No more pressing the "Slope" and "Position" keys.

③ pH measurement of the unknown solution. When the temperature of the test solution is the same as that of the calibration solution, wash the electrode with deionized water, and then dry it with filter paper. Next, insert the electrode into the unknown solution, and read the pH value of the solution after the data on the display is stable.

(3) Measuring the electrical potential. Insert the electrode into the measured solution and then switch the pH/mV button to "mV" mode and read the equilibrium potential value.

(4) Notes

① The jack of the electrode must be kept clean. If not in use, the electrode plug should be inserted into the jack to prevent dust and moisture from entering.

② If the glass electrode bubble cracks or ages, a new electrode should be replaced. New electrodes should be soaked in deionized water for 24h before use.

(Ⅱ) KLT-1 universal coulomb meter

(1) Performance and structure. According to the basic principle of coulomb titration, the KLT-1 universal coulomb meter adopts four detection methods to indicate the ending point, including potentiometric method, amperometric method, equivalence point rising and equivalence point falling, which could automatically control the titration endpoint. The amount of electricity consumed during the titration is then calculated and displayed integrally. The instrument has high analysis accuracy and can be used in environmental monitoring, petrochemical and other fields. The image of KLT-1 universal coulomb meter is shown in Figure 2-8.

Figure 2-8　Photograph of KLT-1 universal coulomb meter (color picture)

(2) Operating rules for KLT-1 universal coulomb meter

① Startup preparation. Install the electrode on the coulomb meter, and then turn it on to warm up for 20 minutes.

② Pre-electrolysis of blank solution

a. Put the electrode into the electrolytic cell, and then connect the working electrode, the counter electrode, and the endpoint indicator electrodes.

Tips: the red clip is connected to the working electrode, the black one is connected to the counter electrode, and the rest are connected to the endpoint indicator electrodes.

b. Add the blank solution into the electrolytic cell and turn on the stirring switch.

c. Press the "current" "up" and "start" buttons successively to adjust the compensation potential.

d. Restore the "polarization potential" button, press the "electrolysis" button, and then set the "work/stop" switch to the "work" mode to start electrolysis.

e. When the coulomb reading is stable, the electric quantity consumed is originated from the electrolysis of the blank solution.

③ Electrolysis of a standard solution

a. Add an appropriate amount of standard solution into the electrolytic cell, press the "electrolysis" button, and then set the "work/stop" switch to the "work" mode to start electrolysis.

b. Read and record the electric quantity consumed until the electric quantity is unchanged and the current pointer is deflected.

④ Electrolysis of sample

a. Add an appropriate amount of the sample solution into the electrolytic cell, press the "electrolysis" button, and then set the "work/stop" switch to the "work" mode to start electrolysis.

b. Read and record the electric quantity consumed until the electric quantity is unchanged and the current pointer is deflected.

⑤ Switching off. Remove the electrodes and power off the machine.

(3) Notes

① The electrolyte should be prepared frequently to keep fresh.

② Clean the electrolytic cell and electrodes before the experiment.

③ The polarity of the electrode must not be wrongly connected. If wrong, clean the electrode immediately.

④ After measuring, release all buttons on the instrument panel and clean the electrode and electrolytic cell with distilled water. Turn off the power and cover the instrument.

(Ⅲ) CHI 660E electrochemical workstation

(1) Performance and structure. The CHI 660E series (Figure 2-9) are universal electrochemical measuring systems. The instrument contains a fast digital signal generator and a direct digital signal synthesizer for high-frequency AC impedance measurement. The instrument features a dual-channel high-speed data acquisition system, a potentio-current signal filter, a multi-stage signal gain, an iR drop compensation circuit, and a potentiostat/ galvanostat. The potential range is $\pm 10V$, while the current range is $\pm 250mA$. The detection limit is less than 10pA and can be used directly for steady-state current measurement on ultramicro electrodes.

Figure 2-9　Photograph of CHI 660E electrochemical station (color picture)

The CHI 600E series integrate almost all commonly used electrochemical measurement techniques. To meet different application requirements and funding conditions, the CHI 600E series are divided into various models. Different models have different electrochemical measurement techniques and functions, while the basic hardware parameters and software performance are the same.

(2) Operating instructions

① Turn on the switch on the back of the instrument.

② Add the detected system (usually a solution of a substance) into a beaker or other suitable container, and then place the required electrodes in the solution.

③ The electrodes generally adopt the three-electrode system, including the working electrode, the counter electrode and the reference electrode, respectively. The connection method is as follows: the green clamp is connected to the working electrode, while the red one is to the counter electrode and the white one to the reference electrode.

④ Double-click on the CHI 660E software on the computer to open the software interface.

⑤ Click on the "Setup" menu in the software to find the "System" command, and select the com 2 port interface.

⑥ Click on the "Setup" menu in the software to find the "Hardware Test" option to perform a system test, and the hardware test results will be displayed on the screen after about one minute.

⑦ To start an electrochemical technique measurement, open "Technique" in the CHI 660E software interface, click the "Cyclic voltammetry" button, set the required potential range and scanning speed, and click the "Run" button. Other techniques are performed the same as CV.

⑧ Click the "Save" button to save the original image, or click "Data list" to copy and save the data.

(3) Notes

① Current overflow should not appear in the detection process. When the software shows that the current is overflowing, stop the experiment and turn off the instrument immediately to check whether there is a short circuit between the electrode system.

② It is strictly forbidden to place the solution above the instrument to avoid splashing the solution into the instrument and causing damage to the mainboard.

③ The instrument should avoid strong vibration or bump.

(Ⅳ) 888 Titrando automatic potentiometric titrator

(1) Performance and structure. The Titrando 888 potentiometric titrator (Figure 2-10) has titration and dosing functions with a built-in EEPROM intelligent chip that automatically identifies burette volume, titrant name, titrant concentration and titer.

(2) Operating instructions

① Powering on. Connect the power supply of the instrument host and the computer cable, double-click the Tiamo icon on the desktop to open the software.

② Confirming the configuration information

a. Click the lower button on the left to switch to the configuration interface, which is divided into 4 sub-windows, namely device, titrant/solution, sensor, and common variables.

b. Check whether the instrument connection in the device is normal.

Figure 2-10   Photograph of Titrando 888 potentiometric titrator (color picture)

c. Check whether the solution is normal. In the titrant / solution window, the required reagent line, the dosing equipment column should show the msb interface where the dosing device is located.

d. Check whether the electrode is connected correctly. In the sensor window, the smart electrode sensor type is in green font, indicating the electrode type. The non-smart electrode sensor has no identification mark.

③ Titrant preparation

a. Clean the metering tube and the metering tube unit. Drain air bubbles from the metering tube and fill it with reagent.

b. Click the manual control icon at the bottom left of the screen to select the appropriate dosing device. Place the titration head in the waste cup, click "Prepare", click the "Start" button, and the preparation process will be carried out.

④ Measurement

a. Click the "Working Platform" button and switch to the working platform interface.

b. In the execution sub-window in the upper-right corner, click the drop-down arrow to the right of the "Method" and select the method to be used for this experiment among the methods that appear. The method you selected appears in the "Method" subwindow on the left.

c. Prepare the sample beaker, place it in a stirrer, and put it on the titration table. Move the electrode and titration head down so that both are immersed below the liquid level. Note: Adjust the electrode height to a certain distance from the stirrer to prevent damage to the electrode.

d. Enter the sample number, sample amount and other information, click "Start". Now the titration starts, and wait for the experiment to be completed.

⑤ Data viewing. Switch to the database interface, Click "File" and "Start". In the dialog box that appears, select the corresponding database and click "Open". You can view the data you need in the selected database.

Ⅲ The Components and Fundamentals of Chromatograph

（Ⅰ） 7820A gas chromatograph

（1） Performance and structure. Gas chromatograph is composed of a gas circuit system, a sampling system, a separation system, a temperature control system, and a detection and recording system. The carrier gas flows out of the high-pressure steel cylinder, after being reduced to the required pressure by the pressure-reducing valve. It is purified by the purifying and drying system and then passes through the pressure stabilizing valve and the rotameter. Subsequently, the carrier gas flows through the vaporization chamber at a stable pressure and at a constant speed, mixes with the vaporized sample, and brings the sample gas into the column for separation. After separation, the components flow into the detector successively with the carrier gas, and then the carrier gas is emptied. The detector converts the change in the concentration or mass of the components into a certain electrical signal, which is amplified and recorded on the recorder to obtain the chromatogram. The photograph of the 7820A gas chromatograph is shown in Figure 2-11.

Figure 2-11　The photograph of gas chromatograph (color picture)

Qualitative analysis can be carried out based on the retention time of each peak obtained on the chromatographic outflow curve, and quantitative analysis can be performed according to the peak area or peak height.

（2） Operation steps

① Machine startup

a. Turn on the computer power and GC switch.

b. Open nitrogen main valve, air (fully automatic air source), and hydrogen generator switch.

c. Open control panel.

d. Set the parameters and wait for the instrument detector and column to heat up, and the instrument shows "Ready".

② Determination

a. Click "Control" and "Single Run". Edit the sample ID, data file name, and vial number.

b. Open the edited method, select the saving path, and click "Start" after registration.

c. After manually injecting the sample, press the "Start" button on the instrument panel immediately to start sample data acquisition.

d. Click "File" to open the completed sample data chromatogram, click "Method" and

"Integration Event" to perform chromatographic editing, data processing, and save the processed data.

③ Powering off

a. After completing the experiment, edit the shutdown method and reduce the column temperature to about 50℃.

b. After the instrument temperature drops, close the GC system. Exit the workstation, power off the instrument, and turn off nitrogen, hydrogen and air.

(3) Notes

① H_2 is dangerous, be sure to check for leaks frequently, and close it immediately when not in use.

② The column should be aged and then connected to the detector to avoid loss and nozzle blockage.

③ The injection port and the glass liner in the inlet should be replaced regularly. The unused injection port and detector should be plugged in.

(II) LC-20AD HPLC

(1) Performance and structure. High-performance liquid chromatograph (HPLC) includes four main parts, a high-pressure system, an injection system, a separation system, and a detection system. LC-20AD system is equipped with two binary solvent pumps, a built-in degasser, a column oven, and an autosampler, and is controlled by LabSolutions workstation software (Figure 2-12).

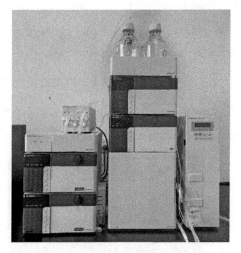

Figure 2-12 The picture of LC-20AD HPLC (color picture)

(2) Operation steps

① Machine startup. Turn on the power supply of the infusion pump, vacuum degasser, system controller, detector system, automatic sampler, and column oven, respectively. Open the computer, and finally click the "LabSolutions" icon to open the workstation interface.

② Editing of analysis methods

a. First open the drain valves on the A and B pumps (rotate 180 degrees in the open direction). Click the "Purge" button on the autosampler panel to automatically empty for 25 minutes.

b. Click the "General" button in the instrument parameter view to set the measurement time, detector wavelength, column oven temperature and flow rate. Save the method and click "Download" to transfer the settings to the instrument.

c. Click the "Turn on System" button. Now the pump starts to work and the column oven turns on. Sampling is only available when the instrument shows as "Ready".

③ Injection

a. When the baseline and pressure are stable, prepare for injection.

b. Select "Single Analysis", enter the data file name and save path. Select the vial number, enter the injection amount, click OK and start the injection.

c. After completing the analysis, wash the column. If the water mobile phase contains salt, it is first necessary to use a mobile phase containing 95% water to clean the column for 20 to 30 minutes. Then, rinse it with pure methanol for 30 min.

④ Powering off. After cleaning, turn off the heating of the pump and the column oven. Turn off the power of each component, and power off the computer.

(3) Notes

① The sealing ring of the high-pressure constant current pump is the most easily worn part, and damage to the sealing ring can cause many failures in the system, which should be maintained and replaced regularly.

② Pumps and detectors must be emptied before injecting the sample.

③ The mobile phase of HPLC grade must be used and filtered by a 0.45μm membrane. The filtered mobile phase must be sufficiently degassed to remove the gas dissolved in it (e.g. O_2). Without degassing, it is easy to produce bubbles, increase baseline noise and decrease the determination sensitivity.

(Ⅲ) BECKMAN P/ACE MDQ capillary electrophoresis system

(1) Performance and structure. Capillary electrophoresis, also known as high-efficiency capillary electrophoresis, is a new type of liquid phase separation technology that uses capillary as the separation channel and high-voltage direct current field as the driving force. The basic structure of the BECKMAN P/ACE MDQ system includes injection, filling/cleaning, current circuit, capillary/temperature control, detection/logging/data processing (Figure 2-13).

Figure 2-13 The picture of BECKMAN P/ACE MDQ capillary electrophoresis system (color picture)

(2) Operation steps

① Machine startup. Turn on the power supply of the BECKMAN P/ACE MDQ capillary electrophoresis system. Open the computer, click the 32 Karat icon and DAD icon to enter the capillary electrophoresis system control interface.

② Sample preparation. Put 0.1mol/L hydrochloric acid, 1mol/L sodium hydroxide, running buffer A, and deionized water into the left buffer reservoir (inlet) and record their respective positions. Then put the buffer bottle with running buffer A and the empty buffer bottle into the buffer reservoir on the right (outlet) and record the corresponding position. Put the buffer bottle containing the sample to be tested into the sample reservoir on the left, and record the corresponding position. Check that the chuck and sample reservoir is installed correctly. Close the reservoir cover and note that the image screen shows that the chuck and reservoir cover is installed. Now you should hear the sound of the refrigerant starting to circulate.

③ Flushing capillary. Click on the pressure area on the control screen and the dialog box appears. Set parameters such as Pressure, Duration, Direction, Pressure Type, Reservoir Positions, etc. Click "OK", move the bottle to the specified position and start rinsing. After rinsing is finished, fill the capillary with the running buffer.

④ Method editing. Go to the main window of 32Karat, and select "Open Offline". Select "File/Method/New", and select "Method/Instrument Setup" to access the method's instrument control and data acquisition modules. Select the "Initial Condition" tab and enter the instrument operating parameters in this dialog box.

⑤ Sequence creation. Select "File/Sequence/New" from the instrument window, open the sequence wizard, and select as required. Click "Finish" and the newly created sequence list appears.

⑥ System running. Check the status of the instrument before running the system. Select "Control/Single Run" from the menu and click the green double arrow on the toolbar of the instrument window to open the "Run Sequence" dialog box.

⑦ Powering off. Turn off the deuterium lamp and click "Load" to bring the reservoir back to its original position. Open the reservoir cover and close the control interface after condensate return. Turn off the switch of CE and computer and cut off the power supply.

(3) Notes

① Only 3 min of rinsing with this buffer is required when running the same buffer. Otherwise, it is necessary to flush with high-purity water and then rinse with buffer.

② After shutting down, the inlet and outlet are sealed with distilled water. If the instrument is not used for a long time, the capillary tube needs to be blown dry and sealed with an empty bottle.

③ High pressure is generated during the operation of the instrument and it is strictly forbidden to open the reservoir cover.

(Ⅳ) Metrohm 861 ion chromatograph

(1) Performance and structure. The Metrohm 861 ion chromatograph (Figure 2-14) is a dual-suppressed ion chromatography instrument with its own conductivity detector. External detectors such as UV/Vis, diode ar-

Figure 2-14　The picture of 861 ion chromatograph (color picture)

ray detector (DAD), voltammetry ampere detector (VA), pulse-ampere detector (PAD) can also be connected. It can also be combined with plasma spectroscopy/mass spectrometry (ICP-AES/MS). Using different ion exchange columns, cations or anions in the specimen can be separated and quantified according to the peak height or peak area of the ion chromatography peak.

(2) Operation steps

① System startup. Double-click the ion chromatography software icon on the desktop to access the operating software.

② Warm-up preparation. Open the system window, click "System、Change" in the system window, and change the system to "anionic system balance". Click "Control、Start measurement" and confirm that there is a negative peak of water after every 10min of the suppressor switching. Warm up for 30 to 60 min until the baseline is stable and click "Control、Stop Measurement".

③ Sample preparation. Standard samples can be directly injected with a disposable syringe, while unknown samples need to be filtered with $0.45\mu m$ membrane filters before sampling. Unknown samples need to be diluted $100\sim1000$ times before sampling to ensure that the concentration is not too high to contaminate the system.

④ Determination. After preheating, click "System、Change" in the window to change the system to "Anionic sample analysis". Click "Control、Start measurement", and input sample information in the pop-up dialog box (name sample as 0 and the standard sample as $1,2,3,\cdots$). Click "OK" to inject the sample into the quantitative ring through the syringe (do not remove the syringe before the next injection). If you want to change the sampling time, click "Method、Property" to enter the sampling time. Repeat the above steps for multiple assays. If there is an expected break of more than one hour, switch the system method to "Anionic System Equilibrium", preventing the suppressor from saturation.

⑤ Powering off the system. After the assay is finished, click "Control、Turn Off Hardware" in the system window to shut down the entire system and finally power off the computer and ion chromatograph.

(3) Notes

① Reagents that have not been used for a long time shall not be stored in the instrument tray, especially hydrochloric acid, which may cause corrosion of instrument components and increase humidity in the instrument.

② If the uncoated capillary is not used for a long time, it should be cleaned with water and then dried in air.

③ If the instrument is not used for a long time, the sample and buffer solution tray must be in the "Load" state before the shutdown.

(Ⅴ) 7890B-5977B GC-MSD gas chromatograph-mass spectrometer

(1) Performance and structure. The Agilent 7890B gas chromatograph (GC) system features accurate temperature control and an accurate sample injection system, as well as a high-performance electronic gas path control (EPC) module for better retention time and peak area reproducibility. The high efficiency source (HES) and InertPlus Extractor EI Source maximize the number of ions that are created and transferred out of the source body and into the quadrupole analyzer. This instrument is widely used in environmental, chemical, petrochemical, food, forensics, pharmaceutical and materials testing (Figure 2-15).

Figure 2-15   The photograph of 7890B-5977B GC-MSD (color picture)

(2) Operating procedures

① Powering on

a. Open the carrier gas cylinder control valve and set the pressure of regulator to 0. 5MPa.

b. Turn on the computer and the 7890B GC, 5977 BMSD power supply in turn. If there is no negative pressure in the MSD vacuum chamber, push the side panel of the vacuum chamber to the right side by hand while turning on the MSD power until the side panel is tightly sucked up. Wait for the instrument self-test to be completed.

c. Double-click the Instrument ♯ 1 icon on the desktop to enter the MSD chemistry workstation.

② Checking the vacuum state. Under the "Instrument Control" interface, click the "View" menu and select "Tune and Vacuum Control" to enter the tuning and vacuum control interface. Select "Vacuum Status" in the "Vacuum" menu to observe the operating status of the vacuum pump. The status should show that the turbo pump speed reaches 100% quickly, otherwise it indicates there is gas leakage in the system. It should check whether the side plates are pressed in place, whether the vent valve is tightened, and whether the columns are connected properly.

③ Checking the water peak and air peak. In the "Instrument Control" interface, click the "View" menu, select "Tune and Vacuum Control", and select "Manul Tune" under "Parameter" to check whether the water peak and air peak meet the requirements.

④ Tuning (generally choose automatic tuning). Click the "View" menu, select "Tune and Vacuum Control", click "Tune", select "Autotune", and the tuning results are automatically printed.

⑤ Configuration editing

a. Click the "Instrument" menu and select "GC Configuration".

b. Under the "Connection" page, enter GC Name, such as "GC 7890B". Enter the configuration of the "7890B" in "Notes", such as "7890B GC with 5977B MSD". Click "Get GC Configuration" to get the configuration of the 7890B.

c. Set ALS, module, column, etc., in turn.

⑥ Method editing

a. Select "Edit Entire Method" in the "Method" menu, select the three items except "Data Analysis", and click "OK".

b. Edit the Notes about the method and click "OK".

c. Set chromatographic parameters such as injector, column mode, valve mode, split or splitless injection, column oven, ALS, schedule, signal and other parameters.

d. Select the mass spectrometry scan mode and set the relevant parameters.

⑦ Data collection and processing

a. Click "Run Method" under the "Method" menu to run the method.

b. Double-click the "Instrument #1 Data Analysis" icon on the desktop to open MSD Data Analysis for data analysis and processing.

⑧ Powering off

a. Under the "Instrument Control" interface, click the "View" menu, select "Tune and Vacuum Control", select "Vent", and click "OK" in the pop-up screen to enter the empty program.

b. Wait until the turbopump speed drops to about 0, and the ion source and quadrupole temperature drops below 100℃, exit the workstation software after about 40min, and turn off the GC, MSD power supply in turn, and finally turn off the carrier gas.

(Ⅵ) 6230 TOF LC/MS

(1) **Performance and structure.** The Agilent 6230 time-of-flight (TOF) LC/MS system is designed to provide superior data quality and advanced analytical capabilities for profiling, identifying, characterizing, and quantifying low molecular weight compounds and biomolecules with confidence. Integrating three core technical innovations—TOF technology, thermal focusing technology and MassHunter workstation software—the 6230 TOF platform is ideally suited for accurate mass analyses of complex samples (Figure 2-16).

(2) **Operation steps**

① Turn on the power supply of the nitrogen generator, and adjust the output pressure to about 0.7MPa (~110psi) after the pressure output is stabilized.

② Turn on the uninterruptible power UPS and confirm that it is in a normal state.

③ Prepare the required mobile phase and flushing solvent of the pump. Check the status of the pipeline link, and confirm that the exhaust gases from the crude vacuum pump and spray chamber are discharged outside the laboratory.

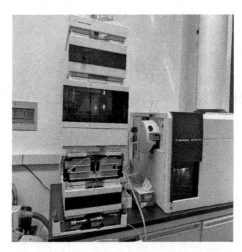

Figure 2-16　The photograph of 6230 TOF LC/MS system（color picture）

④ Turn on the power supply of each module in the LC and the mass spectrometer successively, and wait for the self-test of each module to be completed.

⑤ Power on the computer and network switch.

⑥ Open the MS diagnostic software, double click Pump Down and HV condition, and wait for the condition with TOF high vacuum less than 1.5×10^{-6} and high pressure to complete.

⑦ Double-click the MassHunter acquisition software icon to enter the MassHunter workstation.

⑧ Enter the tuning interface for mass axis calibration.

⑨ Based on the sample tested, establish the LC method. According to the sample structure, select the positive and negative ion modes and the corresponding MS method. Create and run the worklist.

⑩ At the end of the sequence run, clean the pump and column. Open the data analysis software, and conduct data analysis.

⑪ After the sample is completed, stop running the instrument. Close the sampling software, close the various modules of the liquid phase, clean the ion source, and flush the atomizer assembly.

⑫ The mass spectrometer can be in standby mode without running the sample, while it needs to be shut down if it is not used for a long time.

Chapter 3
Instrumental Analysis Experiments

Exp. 1　Spectrophotometric Determination of Iron in Honey with 1,10-Phenanthroline

Objectives

(1) To be familiar with the components and correct use of the spectrometer.

(2) To learn selection of experimental conditions of spectrophotometric analysis.

(3) To learn obtaining the absorption spectrum, plot the working curve and select of maximum wavelength.

Principles

Absorbance photometry is an analytical method based on the selective absorption of light by the molecules of a substance. In absorbance photometric analysis, if the component to be measured is colored and has strong absorbance, it can be directly determined. If the component to be measured is colorless or light in color, it is necessary to select the appropriate reagent to react with the component to produce a colored compound with stronger absorbance, and then measure. The reaction that transforms a colorless or light-colored component into a colored compound (complex, ionic or neutral molecule) is called a chromogenic reaction, and the reagents used are called chromogenic agents.

There are two main types of chromogenic reactions: redox reactions and complexation reactions. In the solution of pH=2-9, the complexation of 1,10-phenanthroline (Phen) and Fe^{2+} converts the very light Fe^{2+} into a stable orange-red complex of Phen and Fe^{2+}. Its formation constant is $K_f = 10^{21.3}$ and molar absorptivity (ε) is $1.1 \times 10^4 \, L/(mol \cdot cm)$. The reaction formula is as follows:

The maximum absorption of the red compound is at a wavelength of 510nm. The method is highly selective. Equivalent to 40-fold the iron content of Sn^{2+}, Al^{3+}, Ca^{2+}, Mg^{2+}, Zn^{2+}, SiO_3^{2-}, 20-fold of Cr^{3+}, Mn^{2+}, V^{5+}, PO_3^{3-}, and 5-fold of Co^{2+}, Cu^{2+} is not found to affect the assay.

Honey is a sweet fluid made by honeybees using the nectar of flowering plants. There are about 320 different varieties of honey, which vary in color, odor and flavor. Honey contains mostly sugar, as well as a mix of amino acids, vitamins, minerals, iron, zinc and antioxidants. In this experiment, the iron content in honey is determined by spectrophotometry using 1,10-phenanthroline as the chromogentic reagent.

Equipment and reagents

(1) **Equipment.** 721 (or 722) spectrophotometer, 10mL pipette, 50mL volumetric flask, 1000mL volumetric flask, 1cm cuvette, porcelain crucible, electric heater, and muffle furnace.

(2) **Reagents**

① 0.0001mol/L iron standard solution: Accurately weigh 0.0482g of ferrous ammonium sulfate hexahydrate, $NH_4(FeSO_4)_2 \cdot 6H_2O$ in a beaker, dissolve it with 30mL of 2mol/L HCl, and transfer it to 1L volumetric flask. Dilute the solution to the mark and shake well (for determination of molar ratio). ② Iron standard solution (containing iron 0.1mg/L): Accurately weigh 0.8634g of $NH_4(FeSO_4)_2 \cdot 6H_2O$ in a beaker, dissolve it with 20mL of HCl (1:1, volume ratio) and a small amount of water, and transfer it to 1L volumetric flask. Dilute the solution to the mark and mix well. ③1,10-phenanthroline solution (1.5g/L freshly prepared aqueous solution). ④100g/L hydroxylamine hydrochloride solution (freshly prepared). ⑤The sodium acetate buffer (1.0mol/L). ⑥Sodium hydroxide solution (0.1mol/L). ⑦Hydrogen chloride solution (1:1). ⑧Honey.

Procedures

(1) **Optimization of experimental conditions**

① Obtaining of absorption curve and selection of measurement wavelength. Pipette 0 and 1.0mL of iron standard solution into two 50mL volumetric flasks (or colorimetric tubes), respectively. To each flask, add 1mL of hydroxylamine hydrochloride solution, 2mL of Phen solution, 5mL of NaAc. Fill each flask to the mark with deionized water, mix well and allow to stand for 10 minutes. Serve the blank reagent (i. e. 0.0mL of iron standard solution) as the reference solution with a 1cm cuvette. Measure the absorbance in 10nm increments in the range of 400 to 600nm. Near the maximum absorption, measure the absorbance in 5nm increments. Plot the graph of absorbance vs. the wavelength of the standards. Select the appropriate wavelength for determining Fe from the absorption curve, generally using the maximum absorption wavelength (λ_{max}).

② Selection of the solution acidity. To each of the 7 flasks (or colorimetric tubes), add 1mL of standard solution, 1mL of hydroxylamine hydrochloride, 2mL of Phen, and mix well. Then, to each flask add 0mL, 2.0mL, 5.0mL, 10.0mL, 15.0mL, 20.0mL, and

30.0mL of a 0.10mol/L NaOH solution with a burette. Dilute each flask to the mark with water, mix well and allow to stand for 10 minutes. Using distilled water as the blank, measure the absorbance of each solution at the selected wavelength using a 1cm cuvette. Meanwhile, measure the pH of each solution with a pH meter. Plot absorbance vs. pH of the solutions to obtain the suitable acidity range.

③ Selection of the amount of developer. To each of the seven 50mL flasks (or colorimetric tubes), add 1mL of iron standard solution, 1mL of hydroxylamine hydrochloride, and mix well. Then add 0.1mL, 0.3mL, 0.5mL, 0.8mL, 1.0mL, 2.0mL, and 4.0mL of Phen and 5.0mL of NaAc solution to each flask. Dilute each flask to the mark with distilled water, mix well, and allow to stand for 10 minutes. Using distilled water as the blank, measure the absorbance of each solution at the selected wavelength using a 1cm cuvette. Using the volume of the taken phenanthroline solution (V) as the x-axis and the absorbance (A) as the y-axis, plot the curve of absorbance (A) versus volume of the developer (V) to determine the most appropriate amount of the developer for measuring iron.

④ Color development time. In a 50mL volumetric flask (or colorimetric tube), add 1mL of iron standard solution, 1mL of hydroxylamine hydrochloride solution, and mix well. Add 2mL of Phen, 5mL of NaAc, dilute to the mark with deionized water and mix well. Measure the absorbance at the selected wavelength using a 1cm cuvette immediately with distilled water as the reference solution. Measure in order the absorbance after the solution is stocked for 5min, 10min, 30min, 60min, and 120min. Plot absorbance on the y-axis, color development time t on the x-axis. Obtain the appropriate time required for the complete reaction of iron with phenanthroline.

⑤ Determination of the molar ratio of phenanthroline to iron. To each of the eight 50mL flasks, pipette 0.0001mol/L iron standard solution in each volumetric flask, add 1mL of 10% hydroxylamine hydrochloride solution and 5mL of 1mol/L NaAc solution. Then add 0.02% phenanthroline solution (about 1×10^{-10} mol/L) 0.5mL, 1.0mL, 2.0mL, 2.5mL, 3.0mL, 3.5mL, 4.0mL, 5.0mL to each flask. Dilute each flask to the mark with water, and mix well. Measure the absorbance of each solution at a wavelength of 510nm using a 2cm cuvette with distilled water as a blank. Finally, plot the absorbance vs. the concentration ratio of phenanthroline to iron (c_R/c_{Fe}, x-axis) and determine the reaction complexation ratio of Fe^{2+} to phenanthroline according to the intersection of the two extension lines on the curve.

(2) Measuring the iron content

① Preparation of the calibration curve. Pipette 10mL of 100g/mL iron standard solution into a 100mL volumetric flask, add 2mL of 2mol/L HCl, dilute to the mark with water, and mix well. This solution contains Fe^{3+} 10μg per mL. To each of the six 50mL volumetric flasks (or colorimetric tubes), add 0mL, 2.0mL, 4.0mL, 6.0mL, 8.0mL, 10.0mL of 10μg/mL iron standard solution with a pipette. Add 1mL of hydroxylamine hydrochloride, 2mL of Phen, 5mL of NaAc solution, and mix well each time after adding a reagent. Then, fill each flask to the mark with water and allow it to stand for 10min after mixing well. Measure the absorbance of each solution at the selected wavelength using a 1cm cuvette with a blank reagent (i. e. 0. 0mL of iron standard solution). Plot the standard curve using the absorbance A as the y-axis, and the concentration of iron as the x-axis. Using the calibration curve, read the absorbance of the corresponding iron concentration and calculate the molar absorptivity of the Fe^{2+}-Phen complex.

② Determination of iron content in honey sample. Accurately weigh 3. 5-4. 0g of honey in a clean porcelain crucible, heat it on the electric heater until it does not smoke, and put it into the muffle furnace at 850℃ for 1. 5h. After cooling down to room temperature, add 2mL (1 : 1) HCl and then boil the solution. Transfer the solution into a 50mL volumetric flask after adding 5～10mL of distilled water. Continue add to the flask 2mL of hydroxylamine hydrochloride, 2mL of Phen, 10mL of 1mol/L NaAc, fill the flask to the mark with water and mix well. Measure the absorbance A and determine the iron content in the sample according to the calibration curve (g/mL).

Data processing

Plot various experimental condition curves and the calibration curve. Calculate the content of iron in honey sample.

Questions

(1) Why should we choose the acidity, the amount of developer and the stability of the colored solution as the test experimental condition?

(2) What is the difference between the absorption curve and the calibration curve? What is the performance of each curve?

(3) What are the roles of hydroxylamine hydrochloride and sodium acetate in this experiment?

(4) How to determine the content of the total iron and ferrous iron in water samples by spectrophotometry? Write out the basic steps.

(5) Can the order of adding reagents be arbitrarily changed when making standard curves and performing other conditions? Why?

Exp. 2   Determination of Nitrite in Food

Objectives

(1) To learn the operation of the spectrophotometer.

(2) To master the method for the determination of NO_2^- in food.

Principles

As a food additive, nitrite maintains the color and aroma of cured meat products and has certain antiseptic properties. But at the same time, it also has a strong carcinogenic effect, and excessive consumption will be harmful to the human body. Therefore, the amount of nitrite added must be strictly controlled in food processing.

In a weak acidic solution, the nitrite reacts with p-aminobenzenesulfonic acid to form a diazo dye, and the resulting diazo compound is coupled with ethylenediamine hydrochloride to form a purple-red azo dye, which can be determined by spectrophotometry. The relevant reactions are as follows:

Equipment and reagents

(1) **Equipment.** 721 spectrophotometer, multi-function food shredder.

(2) **Reagents**

① Saturated borax solution: Weigh 25g of borax ($Na_2B_4O_7 \cdot 10H_2O$) and dissolve it in 500mL of hot water. ②1.0mol/L zinc sulfate solution: Weigh 150g of $ZnSO_4 \cdot 7H_2O$ and dissolve it in 500mL of water. ③150g/L potassium ferrocyanide solution. ④4g/L p-aminobenzenesulfonic acid solution: Weigh 0.4g of p-aminobenzenesulfonic acid and dissolve it in 200g/L hydrochloric acid to prepare a 100mL solution. Store the solution in the dark. ⑤2g/L naphthylethylenediamine hydrochloride solution: Weigh 0.2g of naphthalene ethylenediamine hydrochloride, dissolve it in 100mL of water and store the solution in the dark. ⑥$NaNO_2$ standard solution: Accurately weigh 0.1000g of analytically pure $NaNO_2$ (dried for 24h), dissolve it in water, and transfer it to a 500mL volumetric flask. Dilute to the mark with water and mix well. Accurately pipette 5.0mL of the above stock solution (0.2g/L) into a 100mL volumetric flask, dilute to the mark with water and mix well to obtain a working solution (1μg/mL). ⑦Activated carbon.

Procedures

(1) **Sample pretreatment**

① Meat products (e.g. sausages or ham). Weigh 5g of the ground meat products into a 50mL beaker, add 12.5mL of saturated borax solution and stir the mixture well. Transfer all the samples in the beaker with 150-200mL of hot water (above 70℃) into a 250mL volumetric flask and put it in a boiling water bath for 15min. Take out the flask and precipitate the protein by dropwise addition of 2.5mL of $ZnSO_4$ solution by gentle shaking. After cooling down to room temperature, dilute the flask to the mark with water, mix well and allow it to stand for 10min. Remove the upper layer of fat, and filter the supernatant with filter paper or absorbent cotton. Discard the first 10mL of the filtrate. The filtrate for the measurement should be colorless and transparent.

② Canned fruits, vegetables. Open canned fruits and vegetables, and transfer all the contents to a porcelain plate. Cut the samples into small pieces, mix well, and take out 200g by a point-centered quarter method. Place the sample in a large cup of a food pulverizer with 200mL water, and transfer them to a 500mL beaker for use after the mixture is mashed into a homogenate. Weigh 40g of the homogenate into a 50mL beaker, and wash it into a 250mL volumetric flask with 150mL of hot water (above 70℃) 4～5 times. Add 6mL of saturated borax solution into the flask and mix well. Add continuously 2g of treated activated carbon and mix well again. Then add 2mL of $ZnSO_4$ solution and 2mL of potassium ferrocyanide solution into the flask. Dilute to the mark with water after shaking for 3-5min. Filter the sample with filter paper after well mixing, discard the first 10mL of filtrate, then take out the subsequent 50mL of the filtrate for determination.

(2) Sample determination

① Construction of the standard curve. Accurately pipette the $NaNO_2$ working solution (10g/mL) 0mL, 0.40mL, 0.80mL, 1.20mL, 1.60mL, 2.00mL into six 50mL volumetric flasks, add water of 30mL each, then add 2mL of p-aminobenzenesulfonic acid solution and mix well. After standing for 3min, add 1mL of ethylenediamine hydrochloride solution to each flask, dilute to the mark with water and mix well. After placing for 15min, measure the absorbance of each test solution at a wavelength of 540nm using a 1cm cuvette with a blank reagent as the reference.

② Determination of sample. Accurately pipette 40mL of treated sample filtrate into a 50mL volumetric flask. After diluting to the mark, repeat procedure (2) ① to measure the absorbance of the sample.

Data processing

(1) Plot the standard curve using the concentration of $NaNO_2$ solution as the x-axis and the corresponding absorbance as the y-axis, and obtain the linear regression equation.

(2) Substitute the absorbance value of the sample into the regression equation and calculate the concentration of $NaNO_2$ and the mass fraction of $NaNO_2$ in the sample (expressed in mg/kg).

Notes

(1) Nitrite is easily oxidized to nitrate, and the time and temperature of heating should be carefully controlled when pretreating the sample. In addition, the standard stock solution prepared should not be stored for a long time.

(2) The content of nitrate in the sample is not included in the measurement of this method.

Questions

(1) Check the relevant information to understand the types and dosage of food additives in the production process of Jinhua ham.

(2) Why should the initial 10mL of filtrate be discarded when receiving the filtrate?

Exp. 3   Determination of Anthraquinone and Its Molar Absorptivity by Ultraviolet-Visible Spectroscopy

Objectives

(1) To master the operation skills of UV-Vis spectrophotometer.

(2) To grasp the method for the quantitative determination of anthraquinone.

Principles

Organic compounds with unsaturated structures, such as aromatic compounds, have characteristic absorptions in the ultraviolet region (200-400nm), providing useful information for the identification of organic compounds.

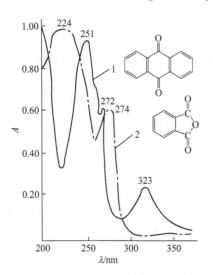

Anthraquinone has a strong absorption at 251nm[$\varepsilon = 4.6 \times 10^4 \mathrm{L/(mol \cdot cm)}$], and a medium-intensity absorption at 323nm [$\varepsilon = 4.7 \times 10^3 \mathrm{L/(mol \cdot cm)}$]. To avoid the interference of the absorption of phthalic anhydride at 251nm, where phthalic anhydride has maximum absorption, the wavelength for the quantitative determination of anthraquinone is therefore set at 323nm (Figure 3-1).

The intensity of light absorbed (A) is directly proportional to the concentration of anthraquinone ($A = \varepsilon b c$), and the content of the sample can be obtained by using the standard curve method. Molar absorptivity (ε) is defined as a measure of a chemical's ability to absorb light at a specified wavelength, which can be found from the slope of the calibration curve.

Figure 3-1   The absorption curves of anthraquinone (curve 1) and phthalic anhydride (curve 2).

Equipment and reagents

(1) **Equipment.** Puxi T9 UV-vis spectrophotometer, a pair of quartz cuvettes with lid (1cm), volumetric flasks (10mL, 100mL, 1000mL).

(2) Reagents

① Anthraquinone (analytical grade). ② Methanol. ③ Unknown sample. ④ 0.1600g/L anthraquinone solution: accurately weigh 0.1600g of anthraquinone in a 100mL beaker, dissolve it with methanol, transfer it to a 1000mL volumetric flask, and dilute to mark with methanol. ⑤ 0.0640g/L anthraquinone solution: pipette 40mL of 0.1600g/L anthraquinone solution into a 100mL volumetric flask, dilute to the mark with methanol, and shake well.

Procedures

(1) **Preparation of anthraquinone standard solutions.** Pipette 1.00mL, 2.00mL, 3.00mL, 4.00mL and 5.00mL of 0.0640g/L anthraquinone standard solution into five separate 10mL measuring flasks, dilute to mark with methanol and shake well.

(2) **Scanning of the absorption curve and selection of measurement wavelengths**

① Place the two cuvettes filled with methanol in the two holders labeled blank and sample accordingly. Do the auto zero scanning from 200 to 350nm.

② Take out the sample cuvette, discard the methanol solution, and replace it with No. 4 anthraquinone standard solution. Scan the absorption spectrum of anthraquinone in the range of 200-350nm, and find the maximum absorption wavelength and the quantitative wavelength.

(3) **Plotting the calibration curve.** Measure the absorbance of each standard solution at 323nm using methanol as your blank.

(4) **Determination of unknown sample.** Pipette 5mL of anthraquinone sample into 10mL measuring flask and dilute to mark with methanol. Measure the absorbance of the sample solution under the same conditions (g/L).

Data processing

(1) Record your data in Table 3-1. Make a standard curve by plotting the absorbance vs. concentration of anthraquinone.

(2) Find the concentration of anthraquinone in the unknown sample from the calibration curve.

(3) Find the molar absorptivity of anthraquinone from the slope of the curve.

Table 3-1 Data record and calculation

λ/nm			Reference solution			
Item	Standard solutions					Sample
No.	1	2	3	4	5	6
V/mL	1.00	2.00	3.00	4.00	5.00	5.00
c/(g/L)						
A						
ε/[L/(mol·cm)]						

Questions

(1) Why do we use 323nm instead of 251nm as the quantitative measurement wavelength for anthraquinone?

(2) What is molar absorptivity? Is it dependent on the concentration of the analyte? Explain why.

Exp. 4　Simultaneous Determination of Chromium and Cobalt by Spectrophotometry

Objectives

(1) To further get familiar with the use of UV-Vis spectrophotometer.

(2) To master the method of mixture analysis.

Principles

Quantitative analysis using UV-Vis spectroscopy is based on the Beer-Lambert Law. When measuring two or more absorbing substances, the absorption of light by each component is additive. Provided that the components do not react with one another, the total absorbance (A) at wavelength λ is the sum of the absorbance of the substances [equations (3-1) and (3-2)],

$$A_\lambda = A_{1\lambda} + A_{2\lambda} + A_{3\lambda} + \cdots \tag{3-1}$$

$$A_\lambda = \varepsilon_{1\lambda} bc_1 + \varepsilon_{2\lambda} bc_2 + \varepsilon_{3\lambda} bc_3 + \cdots \tag{3-2}$$

Where the subscripts 1, 2, and 3 refer to the absorbing components. ε_λ is the molar absorptivity of the substance at wavelength λ. The purpose of this experiment is the simultaneous determination of the concentration of $[Cr(H_2O)_6]^{3+}$ [hexaaquochromium (Ⅲ) ion] and $[Co(H_2O)_6]^{2+}$ [hexaaquocobalt (Ⅱ) ion]. For this two-component mixture, the simultaneous equations (3-3) and (3-4) can be obtained according to absorbance additivity principle:

$$A_{\lambda 1} = \varepsilon_{Cr\lambda 1} bc_{Cr} + \varepsilon_{Co\lambda 1} bc_{Co} \tag{3-3}$$

$$A_{\lambda 2} = \varepsilon_{Cr\lambda 2} bc_{Cr} + \varepsilon_{Co\lambda 2} bc_{Co} \tag{3-4}$$

Among them, $\varepsilon_{Cr\lambda 1}$, $\varepsilon_{Cr\lambda 2}$, $\varepsilon_{Co\lambda 1}$ and $\varepsilon_{Co\lambda 2}$ are the molar absorptivity of Cr and Co at λ_1 and λ_2, which can be calculated by measuring a certain concentration of pure Cr and Co solution at λ_1 and λ_2, respectively. Solving the above system of equations can find c_{Co} and c_{Cr} in the mixture.

In this experiment selection of the wavelength is quite important. The ideal situation is to use a wavelength where one of the components does not absorb at all and the other one absorbs substantially. This would reduce our simultaneous equation problem to one of substitution. Since normally the situation is not so nice then a good choice is one wavelength where most of the absorption is by one species [Co(Ⅱ)] and in the other wavelength, most of the absorption is by the other species [Cr(Ⅲ)].

Equipment and reagents

(1) **Equipment.** Puxi T9 UV-Vis spectrometer, 1cm quartz cell with lid.

(2) **Reagents.** 0.01mol/L Cr(NO$_3$)$_3$ · 9H$_2$O, 0.04mol/L Co(NO$_3$)$_2$ · 6H$_2$O, un-known mixture.

Procedures

(1) **Preparation of standard solutions**

① Pipette 0.010mol/L Cr(NO$_3$)$_3$ · 9H$_2$O solution 1mL, 2mL, 3mL, 4mL in four 100mL volumetric flasks, dilute to the mark with water to obtain 0.0001mol/L, 0.0002mol/L, 0.0003mol/L and 0.0004mol/L Cr(Ⅲ) standard solution.

② Pipette 0.040mol/L Co(NO$_3$)$_2$ · 9H$_2$O solution 1mL, 2mL, 3mL, 4mL in four 100mL volumetric flasks, dilute to the mark with water to obtain 0.0004mol/L, 0.0008mol/L, 0.0012mol/L and 0.0016mol/L Co(Ⅱ) standard solution.

(2) **Selection of λ_1 and λ_2.** In the range of 350-650nm, scan 0.0002mol/L and 0.0008mol/L absorption spectra of Cr(Ⅲ) and Co(Ⅱ) standard solutions, and determine wavelengths λ_1 and λ_2 according to the wavelength selection principle.

(3) **Determination of $\varepsilon_{Cr\lambda1}$, $\varepsilon_{Cr\lambda2}$, $\varepsilon_{Co\lambda1}$ and $\varepsilon_{Co\lambda2}$.** Determine the absorbance of the Cr(Ⅲ) and Co(Ⅱ) standard solutions at λ_1 and λ_2, respectively, and record the data.

(4) **Determination the unkown.** Measure the absorbance of the unknown mixture at both wavelengths.

Data processing

(1) Plot the standard curves using the absorbance values of the two standard solutions at λ_1 and λ_2 as ordinates, and the concentration as the abscissa. Obtain the regression equation and get $\varepsilon_{Cr\lambda1}$, $\varepsilon_{Cr\lambda2}$, $\varepsilon_{Co\lambda1}$ and $\varepsilon_{Co\lambda2}$ from the slopes of the equations.

(2) Calculate the concentration of Cr(Ⅲ) and Co(Ⅱ) in the unknown.

Questions

(1) Briefly describe the principle of UV-visible absorption spectrometry for the determination of binary mixture components.

(2) What other instrumental analysis methods can be used to determine the component content in the mixture of this experiment?

Exp. 5   Measuring Manganese in Steel by Spectrophotometry with Standard Addition Method

Objectives

(1) To understand the principle of spectrometric determination of manganese in steel.

(2) To master the determination principle of the standard addition method.

Principles

Steel is an alloy of iron with small amounts of transition metals such as Mn, Cr, Cu, etc. Steel is digested in hot concentrated (4-5mol/L) nitric acid and analyzed for transition metals by using UV-Vis spectroscopy. Upon digestion, manganese is converted to the colorless Mn^{2+}. An oxidant such as the periodate anion is added to oxidize Mn^{2+} to the familiar deep purple MnO_4^- [Eq. (3-5)]. The concentration of the MnO_4^- is then quantitatively detected by visible spectroscopy. Other metals may interfere with the analysis of Mn by this method and must be removed or masked. The primary constituent of steel is iron which can be masked by the addition of phosphoric acid (H_3PO_4) forming a colorless phosphate complex in the aqueous solution. Other metals such as chromium and cerium can be oxidized with iodate forming potentially interfering colored complexes. To account for these potential interfering ions, the method of standard addition is used.

$$2Mn^{2+} + 5IO_4^- + 3H_2O \Longleftrightarrow 5IO_3^- + 6H^+ + 2MnO_4^- \tag{3-5}$$

Equipment and reagents

(1) **Equipment.** 722 spectrophotometer, 250mL beaker, 10mL pipette, 250mL volumetric flask, 1L volumetric flask, electric stove.

(2) **Reagents.** Steel sample, 4mol/L HNO_3 solution, $(NH_4)_2S_2O_8(s)$, $NaHSO_3(s)$, $Mn(s)$, 85% phosphoric acid, $KIO_4(s)$.

Procedures

(1) **Digestion of steel.** Weigh ～1g steel sample to a 250mL beaker. Record the mass to 0.1mg. Add 50mL of 4mol/L HNO_3 to the beaker and boil gently for a few minutes until the sample is dissolved. Keep your sample covered with a watch glass during the digestion process to prevent loss of material through splattering. Slowly add 1.0g of ammonium peroxydisulfate [$(NH_4)_2S_2O_8$] and boil for 10 to 15 minutes to oxidize any carbon in the sample. If your sample is pink or contains a brown precipitate at this point, add ～0.1g of sodium bisulfite ($NaHSO_3$) and heat for an additional 5 minutes. Allow the solution to cool to room temperature and quantitatively transfer the solution to a 250mL volumetric flask. Dilute with deionized water to the mark. Shake well.

(2) **Preparation of standard Mn solution.** Accurately weigh about 100mg of Mn metal and record the mass of Mn to an accurate 0.1mg. Dissolve Mn metal in 10mL of 4mol/L HNO_3 and boil it for several minutes to remove the generated nitrogen oxides. Quantitatively transfer the solution to a 1L volumetric flask and dilute to the mark with distilled water. Shake well.

(3) **Standard addition.** Transfer a 20mL aliquot of steel solution to a 250 beaker using a volumetric pipette. Add 5mL of 85% phosphoric acid. Add aliquots of standard Mn^{2+} and solid KIO_4 to the beaker according to Table 3-2. Boil each solution for 5 minutes and allow it

to cool to room temperature. Quantitatively transfer the solutions to 50mL volumetric flasks and dilute to the mark. Measure the absorbance of each of the purple solutions using the colorless blank as a reference. Record the absorbance at λ_{max} for the permanganate ion.

Table 3-2   Estimated sample volumes to use when preparing calibration standards

Flask	Sample/mL	H_3PO_4/mL	Std Mn/mL	KIO_4/g
1	0(blank)	5	0.00	0.0
2	20	5	0.00	0.4
3	20	5	2.50	0.4
4	20	5	5.00	0.4
5	20	5	10.00	0.4

Data processing

Plot the standard curve using absorbance as the y-axis and the concentration of added standard Mn^{2+} as the x-axis. The yielded straight line with an x-intercept which is equal to the diluted concentration of Mn from the unknown steel. Calculate the percent by mass of Mn to 0.01% in your steel sample.

Notes

(1) This determination method is suitable for manganese concentrations of 0.2% to 0.5% in steel samples.

(2) If the solution is pink or contains a brown oxide of manganese, add ~0.1g of sodium hydrogen sulfite ($NaHSO_3$) and heat for another 5min.

Questions

(1) Describe the determination principle of the standard addition method.

(2) Discuss what other methods can be used to quantitatively determine the manganese content in steel.

Exp. 6   Dye Colorimetric Determination of Atropine Sulfate Tablets

Objectives

To master the principles and methods of acid dye colorimetry.

Principles

Atropine sulfate $[(C_{17}H_{23}NO_3)_2 \cdot H_2SO_4 \cdot H_2O]$ is a colorless crystal or white crystalline powder. It is odorless, very soluble in water, and soluble in ethanol. Its structural formula is shown in Figure 3-2. Atropine sulfate is an anticholinergic drug, used in the gastrointestinal tract, biliary colic, mydriatic optometry, keratitis, organophosphate pesticide poisoning, septic shock and other syndromes. For the atropine sulfate tablets (0.3mg per piece), the atropine sulfate $[(C_{17}H_{23}NO_3)_2 \cdot H_2SO_4 \cdot H_2O]$ should be within 90.0%-110.0% of the labeled amount.

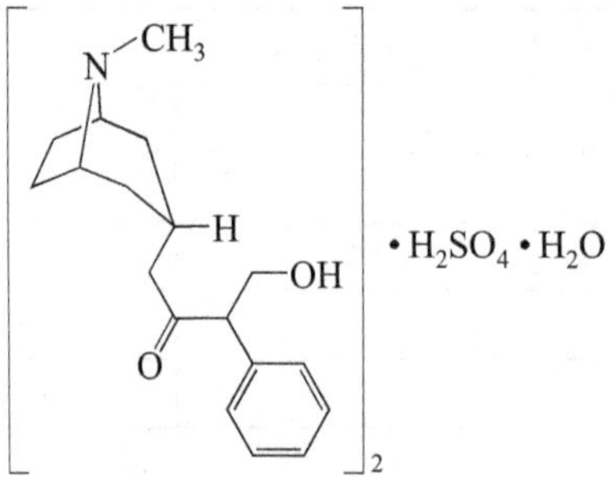

Figure 3-2　The structural formula of atropine sulfate

The content of atropine sulfate tablets can be determined by the acid dye colorimetry. At a certain pH, atropine sulfate can combine with a hydrogen ion to form a cation (BH^+), while the acid dye bromocresol green can dissociate into an anion (In^-). The cation BH^+ and dye anion In^- can be quantitatively combined into a yellow organic complex, yielding an ion pair ($BH^+ \cdot In^-$), which can be quantitatively extracted by trichloromethane. As compared with the reference substance, the labeled content of atropine sulfate tablets can be calculated by measuring the absorbance at the wavelength of 420nm.

Equipment and reagents

(1) **Equipment.** Separatory funnel, UV-Vis spectrophotometer, 100mL beaker, 50mL volumetric flask, 25mL volumetric flask, 100mL volumetric flask.

(2) **Reagents.** Atropine sulfate tablets (commercially available), sulfate reference substance, trichloromethane, bromocresol green, potassium hydrogen phthalate, sodium hydroxide.

Bromocresol green solution: Weigh 50mg of bromocresol green, 1.021g of potassium hydrogen phthalate, and add 6.0mL of 0.2mol/L sodium hydroxide solution to dissolve them. Dilute to 100mL with water and shake well.

Procedures

(1) **Preparation of sample solution.** Accurately weigh 20 pieces of atropine sulfate tablets, finely grind them, and accurately weigh an appropriate amount (about 2.5mg of atropine sulfate) into a 100mL beaker. Add water to dissolve the atropine sulfate and transfer them into a 50mL volumetric flask. Dilute to the mark and shake well. Filter the solution and take the filtrate as a test solution.

(2) **Preparation of reference solution.** Accurately weigh another 25mg of atropine sulfate reference substance, and calibrate in a 25mL volumetric flask. Pipette 5.00mL into a 100mL volumetric flask, and dilute to the mark with water as a reference solution.

(3) Determination of atropine sulfate tablets. Precisely pipette 2.00mL of the test solution and the reference solution, respectively. Transfer them to the separatory funnel with 10mL of chloroform added in advance. Add bromocresol green solution 2.0mL. After 2min of shaking extraction, allow the layers to separate, take the clarified chloroform solution, and measure the absorbance values at a wavelength of 420nm.

Data processing

The formula for calculating the labeled amount of atropine sulfate tablets is as follows:

$$\text{Labeled amount} = \frac{c_R \times \dfrac{A_x}{A_R} \times D \times \overline{W} \times 1.027}{W} \times 100\%$$

Where A_x and A_R are the absorbances of the test sample and the reference respectively, c_R is the concentration of the reference, D is the dilution volume, W is the sample weight, $\overline{W}$ is the average tablet weight, and 1.027 is the molecular weight conversion factor.

Notes

(1) The test sample and the reference product should be operated in parallel, especially in the extraction process, the method, frequency, speed and strength of the oscillation should be consistent.

(2) The separatory funnel must be washed and dried. It should not contain water. A trace amount of water will make chloroform turbid and affect colorimetry. At the same time, the separatory funnel should also use glycerin starch as a lubricant, and can not use soil petrolatum, otherwise, chloroform may dissolve vaseline, causing leakage.

(3) When the acid dye colorimetric method is used to separate the chloroform layer, the head should be "tailed to the tail", and a small amount of the primary filtrate should be discarded. The filtrate of the chloroform layer must be clear and transparent, and must not be mixed with water.

(4) 1.027 in the acid dye colorimetric method is the molecular weight conversion factor of a sample containing 1 molecule of crystalline water atropine sulfate and anhydrous atropine sulfate.

$$\frac{\text{Relative molecular mass of } (C_{17}H_{23}NO_3)_2 \cdot H_2SO_4 \cdot H_2O}{\text{Relative molecular mass of } (C_{17}H_{23}NO_3)_2 \cdot H_2SO_4} = \frac{694.84}{676.84} = 1.027$$

Questions

(1) Why is the content of atropine sulfate drug substance by non-aqueous solution titration method, and the content of atropine sulfate tablets determined by acid dye colorimetry?

(2) What are the main factors affecting the acid dye colorimetric method? What are the most important conditions?

Exp. 7 Determination of Phenol in the Presence of *P*-chlorophenol by Dual-wavelength UV Spectrometry

Objectives

(1) To master the principle of dual-wavelength spectrophotometry.

(2) To master the method of selecting the measurement wavelength and reference wavelength.

Principles

When the two components M and N are in the same solution, it is difficult to obtain the content of individual component by measuring the absorbance at a certain wavelength if their ultraviolet (UV) absorption interfere with each other. The dual-wavelength UV spectrometry can eliminate the interference of one component and determine the content of the other component. As shown in Figure 3-3, λ_2 is the maximum absorption wavelength of the component M. For component M, the absorbance difference (ΔA_M) is the maximum at two wavelengths of λ_1 and λ_2. As a comparison, the absorbance difference at these two wavelengths for N (ΔA_N) is 0. Therefore, measure the absorbance of the mixture at λ_1 and λ_2, and obtain the difference ΔA which is only proportional to the concentration of M as described: $\Delta A = (\varepsilon_M^{\lambda_2} - \varepsilon_M^{\lambda_1})bc_M$.

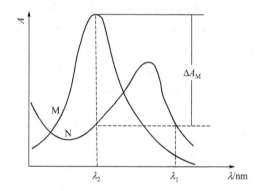

Figure 3-3 Determination principle of the dual-wavelength method

The wavelengths selected for the dual-wavelength method must meet the following requirements. At the two wavelengths, the absorbance of the interference N must be equal, while the absorbance difference of the measured component M should be large enough.

In order to select λ_1 and λ_2, the UV absorption spectra of the individual M and N should be plotted separately on the same coordinate system. Select the λ_2 at or near the maximum absorption peak of component M. Draw a straight line perpendicular to the horizontal axis to obtain a point intersected with the absorption spectrum of interference N. Continue to draw a horizontal line parallel to the horizontal axis at this intersection point and generate one or more intersection points with the spectra of N. Then the wavelength at the intersection point can be used as the reference wavelength λ_1. If there is more than one wavelength can be chosen, select the λ_1 with a large ΔA for the component M.

In this experiment, the absorption spectra of phenol and p-chlorophenol aqueous solution overlap with each other, so that the content of phenol in the mixture can be determined by dual-wavelength UV spectrometry.

Equipment and reagents

(1) **Equipment.** UV-Vis spectrophotometer, two quartz cuvettes (1cm).

(2) **Reagents.** 250mg/L phenol and p-chlorophenol stock solutions: Weigh 25.00mg of phenol and 25.00mg of p-chlorophenol, dissolve them with phenol-free deionized water, and quantitatively transfer them to a 100 mL volumetric flask, respectively. Dilute them to the mark and mix well for further use.

Procedures

(1) **Scanning of absorption spectra of phenol and p-chlorophenol aqueous solution.** Accurately transfer an appropriate amount of the stock solutions and dilute 5 times to 50.0mg/L for the phenol and p-chlorophenol aqueous solution, respectively. Scan the absorption spectra of the above two solutions in the wavelength range of 250~300 nm with phenol-free deionized water as the blank reference in a 1cm quartz cuvette. Plot the two absorption spectra on the same coordinate system to select the appropriate λ_1 and λ_2. Then test the p-chlorophenol aqueous solution to check whether the absorbance is equal at the two wavelengths.

(2) **Construction of standard curve.** Respectively pipette 1.00mL, 2.00mL, 3.00mL, 4.00mL, 5.00mL of 250mg/L phenol stock solution into five 25mL volumetric flasks, dilute to the mark with phenol-free deionized water and shake well. Using a 1cm quartz cuvette. Measure the absorbance of each standard solution with phenol-free distilled water as the blank at the wavelengths of λ_2 and λ_1, respectively.

(3) **Determination of phenol in an unknown sample.** Pipette 5.00mL of unknown sample solution into a 25mL volumetric flask, and dilute to the mark. Under the same conditions, measure the absorbance of the sample at the wavelengths of λ_2 and λ_1, respectively.

Data processing

(1) Plot the absorption spectra of phenol and p-chlorophenol aqueous solution on the same coordinate system, and select the appropriate measurement wavelength λ_2 and reference wavelength λ_1.

(2) Find the difference in absorbance of the standard series solutions ($\Delta A_{\lambda_2 - \lambda_1}$) at two wavelengths. Plot the standard curve with $\Delta A_{\lambda_2 - \lambda_1}$ as the ordinate, while the concentration of phenol aqueous solution as the abscissa. Obtain the linear equation. Substitute $\Delta A_{\lambda_2 - \lambda_1}$ value of the unknown sample solution into the regression equation to calculate the corresponding phenol content. Then calculate the phenol concentration in the unknown sample solution.

Questions

(1) What are the similarities and differences between dual-wavelength instruments and single-wavelength instruments?

(2) What are the two requirements for the wavelength selected in this experiment?

Exp. 8　Quantitative Analysis of Vitamin B$_2$ by a Fluorometric Method

Objectives

(1) To be familiar with the composition and use of fluorescence spectrometers.

(2) To master the method of determining vitamin B$_2$ content by fluorescence method.

Principles

Fluorescence is a form of photoluminescence. A photodetector can detect two processes of fluorescent molecules. One is the excitation process, the other is the fluorescence emission process. The emission process of fluorescence is the transition of electrons from the lowest energy excited singlet state to the ground state. In a dilute solution, the fluorescence emission intensity is proportional to the concentration of the substance, and thus a quantitative analysis of the substance can be performed.

Generally speaking, substances with strong fluorescence often have the following characteristics in their molecular structure: ①It has a large conjugate bond system; ②It has a rigid plane structure; ③ The conjugate system contains an electron-donating substitute group; ④ Its lowest electron excitation singlet state is π to π^* type. Vitamin B$_2$, also known as riboflavin, is an orange-yellow, odorless needle-like crystal. Its structural formula is shown in Figure 3-4. It has a rigid plane and a good conjugate structure, emits green fluorescence under blue light irradiation at 430~440nm in a weakly acidic or neutral aqueous solution, and its maximum emission wavelength is about 525nm.

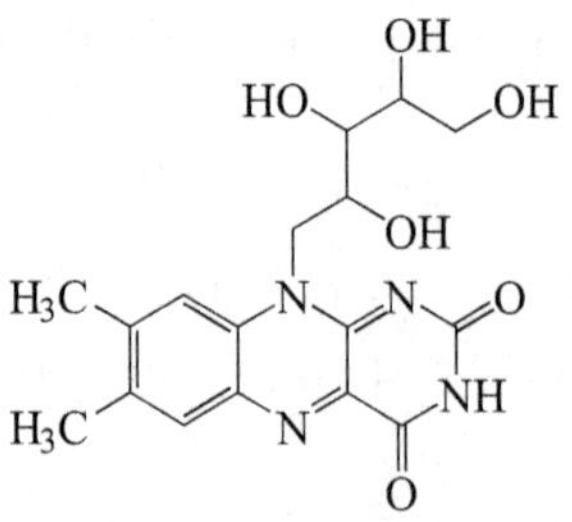

Figure 3-4　The molecular structure of vitamin B$_2$

At lower concentrations, the fluorescence emission intensity I of a substance is proportional to its concentration c. In this experiment, the content of vitamin B$_2$ in drugs is analyzed using the fluorometric method.

Equipment and reagents

(1) **Equipment.** LS 55 fluorescence spectrometer, quartz cuvette, 5mL pipette and 50mL volumetric flask.

(2) **Reagents.** Vitamin B_2 stock solution (10.0mg/L), vitamin B_2 tablets.

Procedures

(1) **Preparation of standard vitamin B_2 solutions.** Pipette 0.50mL, 1.00mL, 1.50mL, 2.00mL and 2.50mL of vitamin B_2 stock solution (10.0mg/L) separately into five volumetric flasks (50.0mL), dilute them to the mark, and shake well for further use.

(2) **Scanning excitation and emission spectra.** According to the operation manual of the LS 55 fluorometer, set the excitation wavelength at 425nm and record the emission spectrum in the wavelength range of 500 ~ 600nm. Set the emission wavelength at 525nm and then record the excitation spectrum in the range of 400~500nm. Find the maximum excitation and emission wavelength on the excitation and emission spectra, respectively.

(3) **Fluorimetric determination of standard solutions.** At the maximum excitation wavelength, scan the fluorescence emission spectra of each standard solution and record the fluorescence intensity at the maximum emission wavelength.

(4) **Real sample analysis**

① Accurately weigh and grind 20 tablets of vitamin B_2, weigh 0.05-0.08g of ground powder, and add 50mL of 1% HAc solution. After heating and dissolving in a water bath for 30 minutes, quantitatively transfer the resulting solution to a 250 mL volumetric flask, dilute to the mark and mix well. Transfer the as-obtained 1.00mL solution into a 50mL volumetric flask, dilute with distilled water to the mark, and mix well.

② Determine the fluorescence intensity of the tablet solution under the same conditions.

Data processing

(1) Using the fluorescence intensity value of each standard solution as the y-axis and its concentration as the x-axis, obtain the standard curve and the linear fitting equation.

(2) Substitute the fluorescence intensity of the sample into the regression equation, calculate the concentration of vitamin B_2, and convert it to the content in the original vitamin B_2 tablet.

Questions

(1) Describe the similarities and differences between a spectrofluorometer and a spectrophotometer.

(2) How to obtain the fluorescence excitation spectra and emission spectra of a substance?

Exp. 9 Determination of O-hydroxybenzoic Acid and M-hydroxybenzoic Acid by Fluorescence Analysis

Objectives

(1) To be further familiar with the operation skills of the fluorescence spectrometer.

(2) To master the multi-component determination methods in fluorescence analysis.

Principles

O-hydroxybenzoic acid (also known as salicylic acid) and m-hydroxybenzoic acid have the same molecular composition and both contain a fluorescence-emitting benzene ring but have different fluorescence properties due to the different positions of their substituents. In the alkaline solution of pH$=$12, both will emit fluorescence under the excitation of ultraviolet light near 310nm. While in the solution of pH$=$5.5, m-hydroxybenzoic acid does not emit fluorescence, and o-hydroxybenzoic acid emits strong fluorescence due to the formation of hydrogen bonds in the molecule which increases the molecular rigidity. Using this property, the content of o-hydroxybenzoic acid in the mixture can be determined at pH $=$ 5.5. Take the same quality of mixture solution, measure the fluorescence intensity of pH$=$12, and subtract the fluorescence intensity of o-hydroxybenzoic acid at pH$=$5.5, the content of m-hydroxybenzoic acid can thus be determined.

Equipment and reagents

(1) **Equipment.** LS 55 fluorescence spectrophotometer, 10mL colorimetric tube, pipette.

(2) **Reagents.** 60μg/mL o-hydroxybenzoic acid standard solution, 60μg/mL m-hydroxybenzoic acid standard solution, 0.1mol/L NaOH aqueous solution. HAc-NaAc buffer solution with pH$=$5.5: 47g NaAc and 6g glacial acetic acid are dissolved in water and diluted to 1L.

Procedures

(1) **Preparation of a series of standard solutions**

① Pipette 0.40mL, 0.80mL, 1.20mL, 1.60mL, and 2.00mL of o-hydroxybenzoic acid standard solution into numbered 10mL colorimetric tubes, add 1.0mL of HAc-NaAc buffer solution with pH$=$5.5, dilute to the mark with deionized water, and shake well.

② Pipette 0.40mL, 0.80mL, 1.20mL, 1.60mL, and 2.00mL m-hydroxybenzoic acid standard solution into numbered 10mL colorimetric tubes, and add 1.2mL of 0.1mol/L of NaOH aqueous, dilute to the mark with deionized water and mix well.

③ In each of the two colorimetric tubes (10mL), pipette 2.00mL of the unknown solution, respectively. Add 1.0mL of HAc-NaAc buffer solution with pH$=$5.5 in one of the tubes, and add 1.2mL 0.1mol/L of NaOH solution into the other tube, dilute to the mark with deionized water and mix well.

(2) **Scanning fluorescence excitation and emission spectra.** Set the emission wavelength at 400nm, and scan the excitation wavelength in the range of 250-350nm to obtain the excitation spectrum and the maximum fluorescence excitation wavelength λ_{ex}^{max} of o-hydroxybenzoic acid and m-hydroxybenzoic acid solution (using the third standard solution). Set the λ_{ex} and scan the emission wavelength at 350-500nm to obtain the emission spectrum and the maximum emission wavelength λ_{em}^{max} of both solutions.

(3) **Fluorescence intensity determination.** According to the above excitation spectrum and emission spectrum scanning results, find a pair of wavelengths (λ_{ex} and λ_{em}), ensure it has a high sensitivity to both components. Determine the fluorescence intensity of each standard solution and unknown solution undertheabove λ_{ex} and λ_{em}.

Data processing

(1) Using I_f of each standard solution as the y-axis and the concentration of o-hydroxybenzoic acid or m-hydroxybenzoic acid as the x-axis, plot the standard curves and get the linear equations.

(2) Substitute the fluorescence intensity of the unknown solution at pH=5.5 into the linear regression equation of o-hydroxybenzoic acid, and calculate the concentration of o-hydroxybenzoic acid in the unknown solution.

(3) Substitute the difference of fluorescence intensity at pH=12 and pH=5.5 into the regression equation of m-hydroxybenzoic acid, and calculate the concentration of m-hydroxybenzoic acid in the unknown solution.

Questions

(1) What do λ_{ex}^{max} and λ_{em}^{max} represent?

(2) Summerize factors that affect the fluorescence intensity of the substance.

Exp. 10 Spectrofluorimetric Determination of Aluminum Ions via Complexation with 8-Hydroxyquinoline

Objectives

(1) To master the basic principles for the determination of aluminum ions by direct fluorescence spectroscopy.

(2) To be familiar with basic operations such as fluorescence spectrometry and solvent extraction.

Principles

Tea tree is an aluminum-accumulating plant, and the aluminum content in tea has an important influence on the growth and quality of tea plants. At present, the commonly used methods for determining aluminum are spectrometry and fluorescence spectrometry, of which fluorescence spectrophotometry has high sensitivity and is more widely used.

Aluminum ion has no fluorescence and cannot be directly determined by fluorescence spectroscopy. However, the aluminum ion is known to form a highly fluorescent complex with 8-hydroxyquinoline (8-hydroxyquinoline aluminum, $C_{27}H_{18}AlN_3O_3$). Its molecular structure is shown in Figure 3-5. The complex is fat-soluble which can be effectively removed from chloroform extracted from the aqueous phase. The extract is measured by fluorescence with the excitation and emission wavelength at 390nm and 510nm, respectively. Because the fluorescence emission intensity of the complex is proportional to its concentration, a direct fluorescence spectrometry for the determination of aluminum ions can be established.

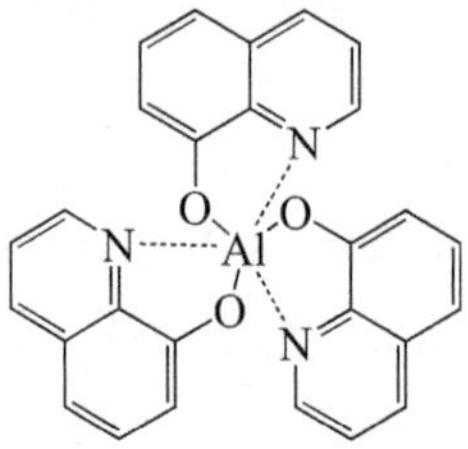

Figure 3-5   Molecular structure of 8-hydroxyquinoline aluminum

Equipment and reagents

(1) **Equipment.** LS 55 fluorescence spectrometer, quartz cuvette, separatory funnel (125mL), pipette, volumetric flask, digestion tank, oven.

(2) **Reagents**

① Dissolve 1.760g of potassium aluminum sulfate [$Al_2(SO_4)_3 \cdot K_2SO_4 \cdot 24H_2O$] in 20mL of water, add 1 : 1 sulfuric acid to the solution, transfer it to a 100mL volumetric flask, dilute to the mark with distilled water and mix well. Accurately transfer the resulting solution of 2.00mL to a 1000mL volumetric flask, dilute to the mark with distilled water and mix well to obtain 2.0μg/mL aluminum ion stock solution. ②Dissolve 2g of 8-hydroxyquinoline in 6mL of glacial acetic acid and make it up to 100mL with distilled water to obtain an 8-hydroxyquinoline solution (2%). ③Prepare a buffer solution containing 200g ammonium acetate and 70ml concentrated ammonia per liter. ④Nitric acid, perchloric acid, 6mol/L ammonia, chloroform, tea sample.

Procedures

(1) **Preparation of a series of standard solutions.** To each of the 6 volumetric flasks (50mL), add 0mL, 10.0mL, 20.0mL, 30.0mL, 40.0mL, and 50.0mL of aluminum ion stock solution, respectively. Dilute each flask to the mark with deionized water, and mix well.

(2) **Formation and extraction of fluorescent complexes.** In each of the six 125mL separatory funnels, add 45mL of water and 5.0mL of the above standard solutions. Then, add

2mL of 8-hydroxyquinoline solution and 2mL of buffer solution to each funnel along the funnel wall. After shaking for 5min, extract twice with 10mL of chloroform each time. Drain the organic phase through dry absorbent cotton into a 50mL volumetric flask, and rinse the absorbent cotton with a small amount of chloroform. Receive the rinse solution into the volumetric flask, dilute each solution to the mark with chloroform and mix well.

(3) **Scanning excitation and emission spectra.** Set the excitation wavelength at 390nm, scan the emission spectrum between 450 and 600nm. Set the emission wavelength at 510nm, and scan the excitation spectrum in the wavelength range of 330-460nm. Find the maximum excitation and emission wavelength on the excitation spectrum and the emission spectrum, respectively.

(4) **Measurement of standard solutions.** At the maximum excitation wavelength, scan the fluorescence spectrum of each standard solution in the wavelength range of 450-600nm. Record the fluorescence intensity at the maximum emission wavelength. Measure each solution three times and take the average.

(5) **Determination of unknown samples.**

① Accurately weigh 0.5000g of tea samples (two parts each), put them into a digestion tube, add HNO_3-$HClO_4$ (4 : 1), and digest it at 400℃ until the solution is transparent. Transfer the transparent solution into a 50mL volumetric flask and rinse the digestion tube several times with deionized water. Make up the volume to 50mL. Pipette 5mL of the solution into a 25mL volumetric flask and dilute to the mark with deionized water.

② Pipette 5mL of the pretreated tea sample solution to a 125mL separatory funnel, and adjust the pH to neutral using 6mol/L ammonia. Continue treating the sample as step (2).

③ Measure the fluorescence intensity of the sample according to step (4).

Data processing

(1) Using the fluorescence intensity of the series standard solution as the y-axis, and the concentration as the x-axis, obtain a fitting equation using linear regression.

(2) According to the fluorescence intensity of the tea sample, calculate the aluminum content by the regression equation.

Questions

(1) Why is the chloroform extract filtered with dry absorbent cotton?

(2) Can the separatory funnel stopcock be treated with vaseline? Why?

Exp. 11   Fluorometric Determination of Acetylsalicylic Acid in an Aspirin Tablet

Objectives

(1) To be familiar with the use of fluorescence photometer.

(2) To master the treatment methods of solid samples.

Principles

Aspirin tablet is a common pain reliever, in which acetylsalicylic acid is the main ingredient. It also contains other ingredients, such as adhesives and buffers. In this experiment, the aspirin tablet is dissolved in water and converted to salicylate ions by the addition of sodium hydroxide (Figure 3-6). The salicylate ion can be excited at about 310nm and emit strong fluorescence at about 400nm. A series of standard solutions of the salicylate ion are prepared, and then the fluorescence intensity of the standard and sample solutions are measured. The standard curve method is used to determine the concentration of salicylate ions in the sample solutions and further calculate the percentage of acetylsalicylic acid in the aspirin.

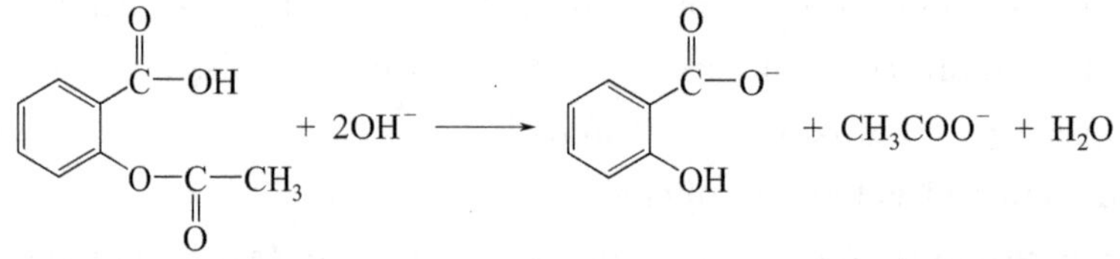

$$\text{Acetylsalicylic acid} + 2OH^- \longrightarrow \text{Salicylate ion} + CH_3COO^- + H_2O$$

Figure 3-6 Conversion of acetylsalicylic acid to salicylate ion under alkaline conditions

Equipment and reagents

(1) **Equipment.** LS 55 fluorometer, 1cm quartz cuvettes.

(2) **Reagents.** Aspirin tablet, $100\mu g/mL$ stock solution of salicylic acid, 1mol/L sodium hydroxide solution.

Procedures

(1) **Sample preparation**

① Place the aspirin tablet in a clean mortar and grind it into powder. Weigh 0.005g of the powder (accurate to 0.001mg) into a 100mL beaker and dissolve it in hot water (80℃). Fold a piece of filter paper and place it in a glass funnel. Place the funnel on top of a 100mL volumetric flask. Allow the solution flowing through the funnel into the volumetric flask. Dilute the solution to the mark with deionized water and shake well.

② Accurately transfer 1.00mL of the previous solution to a 100mL volumetric flask, dilute with deionized water to the mark and shake well.

③ In three 50mL volumetric flasks, pipette 2.00mL of sodium hydroxide solution into each volumetric flask and add 5.00mL of diluted tablet solution. Dilute the three volumetric flasks to the mark with deionized water and shake well.

(2) **Standard solution preparation**

① Exactly prepare $1\mu g/mL$ solution of salicylic acid from the $100\mu g/mL$ stock solution.

② In five 50mL volumetric flasks, pipette 2.00mL of sodium hydroxide solution into each volumetric flask, and then add 1.00mL, 2.00mL, 3.00mL, 4.00mL, 5.00mL of

$1\mu g/mL$ salicylic acid solution into the flasks, respectively. Dilute to the mark using deionized water.

(3) Scanning the excitation and emission spectrum. Set the excitation wavelength at 310nm. Take the No. 1 standard solution as the test solution, and scan the emission spectrum at 320-600nm. With the obtained maximum emission wavelength, scan the excitation spectrum in the range of 200-380nm. Find the maximum excitation wavelength and maximum emission wavelength of salicylic acid.

(4) Measuring unknown

① At the best excitation and emission wavelengths, measure the fluorescence intensity of all standard solutions sequentially.

② Under the same conditions, measure the fluorescence intensity of the sample solution.

Data processing

(1) Prepare a working curve by plotting the fluorescence intensities of salicylate ion standard solutions as a function of their concentrations.

(2) According to the fluorescence intensity of unknown samples, calculate the concentration of salicylate ion in diluted solutions by the regression equation.

(3) Calculate the mass of acetylsalicylic acid (mg) in the weighed tablet powder according to the dilution ratio.

(4) Calculate the percentage ratio of acetylsalicylic acid in the tablet according to the mass of acetylsalicylic acid in the tablet powder. Report the average and standard deviation of acetylsalicylic acid in the tablet.

Questions

(1) Draw a block diagram of the spectrofluorometer.

(2) Consult literature and discuss other instrumental analysis methods of acetylsalicylic acid.

Exp. 12   Determination of Benzoic Acid and Ethyl Acetate by Infrared Spectroscopy

Objectives

(1) To learn the basic principle and instrument structure of infrared spectroscopy.

(2) To understand the application range of infrared spectroscopy.

(3) To master the sample preparation methods of solid and liquid.

Principles

The infrared spectrum reflects the vibration of molecules. When a substance is irradiated with a certain frequency of infrared light, if the vibration frequency of a group in the molecule of the substance is the same, the substance can absorb such infrared light, causing the molecule to transition from the vibrational ground state to the excited state. When infrared light of different frequencies passes through the substance to be measured, absorption of different strengths will occur.

Different compounds have their characteristic infrared spectra, which can be used to analyze the structure of the substance. It is a process of light absorption, substances can also be quantitatively analyzed according to Beer's law.

Equipment and reagents

(1) **Equipment.** Nicolet iS50 Fourier transform infrared spectrometer (FTIR), hydraulic tablet press, infrared drying lamp.

(2) **Reagents.** Anhydrous ethanol, ethyl acetate, benzoic acid, KBr (spectrum pure), agate mortar, salt flakes, unknown sample.

Procedures

(1) **Determination of solid benzoic acid.** Take about 1mg of benzoic acid sample in a clean agate mortar, add about 100mg KBr powder, and grind it into a fine powder with a particle size of about $2\mu m$ under the infrared lamp. Transfer the powder into the pellet die, set the die to the hydraulic pellet press, and maintain a pressure of 16MPa for 2min. Degas and loosen the pressure. Take out the die and obtain a semitransparent KBr pellet with a diameter of 13mm. Put the salt pellet in the sample holder for infrared spectrometric determination.

(2) **Determination of liquid ethyl acetate.** Add a drop of ethyl acetate sample to a clean and polished NaCl or KBr pellet, press another salt pellet onto it, and place the pellets in the sample holder for infrared spectrometric determination.

(3) **Determination of unknown.** According to the unknown material provided, select the sample preparation method and scan the infrared spectrum of the sample.

Data processing

(1) Attribute the characteristic bands of benzoic acid and ethyl acetate.

(2) Speculate on the possible structure of the unknown.

Notes

(1) Grind solid samples under infrared lamp to prevent moisture absorption.

(2) Salt tablets should be kept dry and transparent. They should be polished with absolute ethanol and talc (under infrared lamp) before each measurement, and should not be washed with water.

Questions

(1) What are the similarities and differences between an infrared absorption spectrome-

ter and a UV/Vis absorption spectrometer?

(2) NaCl and KBr are generally used as dispersants when testing infrared spectroscopy. What is the applicable wavenumber range for each?

Exp. 13   Determination of Water Hardness of Tap Water by Flame Atomic Absorption Spectrometry

Objectives

(1) To be familiar with the structure and operation skills of atomic absorption spectrometers.

(2) To master the methods of atomic absorption spectroscopy for elemental quantitative analysis (standard curve method and standard addition method).

(3) To master the calculation of determination sensitivity by atomic absorption spectrometry.

Principles

Water hardness is defined as the total concentration of alkaline earth metal ions in water. Because the concentrations of Ca^{2+} and Mg^{2+} are usually much higher than those of other alkaline earth ions, the total hardness of water can be equal to the total concentration of calcium and magnesium ions in water. Individual hardness refers to the individual concentration of each alkaline earth ion. The total hardness is often expressed as the amount of $CaCO_3$ (mg) per liter in water. One liter of water with a hardness of more than 60 mg of $CaCO_3$ is considered hard water. Water hardness can be determined by many methods, such as EDTA titration, atomic absorption spectroscopy (AAS) and atomic emission spectroscopy (AES), etc.

AAS has three main atomization systems including flame atomization system, graphite furnace atomization system and low-temperature atomization system. Among them, flame atomization is widely used due to its simple operation, fast analysis, low cost and high sensitivity to most elements. In this experiment, flame AAS is used to determine the concentrations of Ca^{2+} and Mg^{2+} in tap water.

Flame AAS is a quantitative analysis method based on the absorbance of the ground state atomic vapor (coming from the flame atomizer) of the measuring element to the characteristic spectral line of the same element radiated from the light source. Under certain conditions, the absorbance is proportional to the concentration of the element to be measured in the test solution, that is, $A = Kc$. The content of elements can be calculated using A-c standard curve method or standard addition method.

When using the standard curve method, a series of standard solutions of the measuring elements are prepared first, and their absorbances (A) are measured respectively. A is plotted against c, and the standard curve is obtained by linear regression. Under the same measurement conditions, the absorbance of the solution to be tested (A_x) is measured, and the concentration c_x of the analyte can be obtained from the linear equation.

When using the standard addition method, 5 equal amounts of the sample solution are transferred into five volumetric flasks, respectively. In the second to fourth volumetric flasks, a series of standard solutions are added, and the absorbance is determined after the volume is determined. Plot the absorbance curve of the measured element, extrapolate this curve, and the intersection point with the concentration coordinate is the content of the element to be measured in the diluted sample.

In flame AAS, the sensitivity is defined as the concentration of analyte that produces 1% absorbance signal ($A = 0.0044$). According to the definition, the sensitivity of an element is:

$$S = \frac{c \times 0.0044}{A}$$

Where c is the concentration of the analyte, and A is the absorbance. Obviously, the smaller the S, the higher the sensitivity of the elemental determination.

Equipment and reagents

(1) **Equipment.** GGX-810 atomic absorption spectrophotometer, Ca and Mg hollow cathode lamps (measure the absorbance of Ca and Mg at 422.7nm and 285.2nm, respectively), air compressor, and acetylene cylinder.

(2) **Reagents.** 100mg/L Ca standard solution, 50mg/L Mg standard solution, tap water.

Procedures

(1) **Setting of instrument parameters.** Set instrument parameters according to optimal instrument conditions.

(2) **Preparation of standard series**

① Ca standard solution series: prepare 2.00mg/L, 4.00mg/L, 6.00mg/L, 8.00mg/L and 10.00mg/L of Ca standard solutions in five 50mL volumetric flasks.

② Mg standard solution series: prepare 0.25mg/L, 0.50mg/L, 0.75mg/L, 1.00mg/L and 1.25mg/L of Mg standard solutions in five 50mL volumetric flasks.

(3) **Determination of water hardness**

① Under the selected working conditions, measure the absorbance of Ca and Mg standard solutions with ultra-pure water as blank.

② Measure the absorbance of Ca and Mg ions in tap water with ultra-pure water as blank.

(4) **Shutting down.** After the experiment, spray the atomization system with ultrapure water for 2min and shut down the instrument according to the operation programs.

Data processing

(1) Tabulate the experimental data of Ca and Mg ions, and plot the standard curves

with the concentration of metal elements as the abscissa and the measured absorbance value as the ordinate.

(2) Calculate the concentration of Ca and Mg ions in tap water and the hardness of water, and judge the hardness degree of water.

Questions

(1) Describe the basic principle of AAS.

(2) Why is a hollow cathode lamp of the measuring element used as the light source for AAS?

Exp. 14   Determination of Cu and Pb in Brass by Flame Atomic Absorbance Spectroscopy

Objectives

(1) To be further familiar with the operation of atomic absorbance spectrophotometer.

(2) To understand the pretreatment process of brass.

Principles

Brass is an alloy made primarily of copper and zinc. The proportions of the copper and zinc are varied to yield many different kinds of brass. The amount of copper may range from 55% to 95% by weight, basic modern brass is 67% copper and 33% zinc.

Lead is commonly added to the brass at a concentration of around 2%. The lead addition improves the machinability of brass. In this experiment, copper and lead content in brass are determined by flame atomic absorbance spectroscopy (AAS) using the standard curve method.

Equipment and reagents

(1) **Equipment.** GGX-810 atomic absorption spectrophotometer, Pb and Cu hollow cathode lamps (measure the absorbance of Pb and Cu at 283.3nm and 324.8nm, respectively), air compressor, and acetylene cylinder.

(2) **Reagents.** Concentrated HNO_3, lead shot, Cu foil, brass.

Procedures

(1) **Preparation of Pb stock solution.** Accurately weigh about 0.5g of Pb shot to a 150mL beaker (do not dry the lead in an oven). In a fume hood add 20mL of distilled water and 20mL of concentrated nitric acid. Cover with a watch glass and boil gently until the metal has dissolved and the solution is colorless. If a white precipitate forms, cool the solution and add 20mL of water. Remove from the heat and allow to cool slightly. Rinse the watch glass and the sides of the beaker with a small amount of distilled water and boil for an additional 10

minutes. Allow the solution to cool and quantitatively transfer it to a 1000mL volumetric flask. The amount of lead added should yield a final concentration of lead of 500mg/L.

(2) **Preparation of Cu stock solution.** Weigh a~0.5g sample of copper foil and digest it with 50mL of dilute HNO_3 (3mol/L) in a fume hood. Gently boil the solution for 10 minutes after the last of the copper has been dissolved and the solution is blue and transparent. Keep the beaker covered with a watch glass and do not let the solution evaporate to dryness. Cool the sample to room temperature and quantitatively transfer the solution to a 1000mL volumetric flask. Dilute to the mark with distilled water.

(3) **Preparation of brass sample stock solution.** Weigh a 1.0g sample of brass and add it to a 150mL beaker. Add 10mL of water followed by 15mL of concentrated nitric acid in a fume hood. After the evolution of gas slows, add 10mL of water. Heat the solution and continue to carefully boil the covered solution for an additional 20 minutes after all of the brass has dissolved. The solution will be reduced in volume but do not allow it to evaporate to dryness. Allow the solution to cool to room temperature. A fluffy white precipitate may be visible in the blue solution. The precipitate is a hydrated stannic oxide (H_2SnO_3). Carefully filter the solution through a cone of filter paper into a clean beaker. Quantitatively transfer the solution to a 500mL volumetric flask. Dilute to the mark with deionized water and shake well.

(4) **Sample analysis**

① Prepare Pb and Cu standard solutions (between 0 and 50mg/L) using their stock solutions.

② Measure the absorbance of each calibration solution.

③ Measure the absorbance of copper and lead in unknowns. Adjust the dilution of the brass unknown so that the concentrations of Cu and Pb fall within the linear range of the standard curve.

Data processing

(1) Construct a calibration curve by plotting the average absorbance vs. the concentration of metals. Include error bars for each point (standard deviations).

(2) Calculate the concentration of Pb and Cu in your unknown sample using the calibration curves. And with 95% confidence, report the confidence interval for the content of Pb and Cu in the sample (accurate to 0.01%).

Notes

(1) Record the mass of copper to the nearest 0.1mg.

(2) Handle all the weighing samples with gloves to avoid fingerprints and do not dry them in an oven before weighing.

Questions

(1) If there is a fluffy white precipitate in step (3), how to gravimetrically determine

tin in the sample?

(2) Find a research paper that uses other methods to determine Cu content in brass and explain how this method was used in the study.

Exp. 15　Determination of Cadmium Content in Soil by Graphite Furnace Atomic Absorption Spectrometry

Objectives

(1) To use the graphite furnace atomic absorption spectrometer.

(2) To understand the pretreatment technology for cadmium determination in soil.

(3) To master the two quantitative methods of cadmium determination (standard curve method and standard addition method).

Principles

The atomization systems commonly used in atomic absorption spectroscopy (AAS) are flame atomization system, graphite furnace atomization system and low temperature atomization system. Different types of atomization systems directly affect the sensitivity, detection limits, precision, and linearity range of elemental analysis. Among them, flame atomization is widely used due to its simple operation, fast analysis, low cost and high sensitivity to most elements. However, flame atomization requires a large volume of sample solution and has low atomization efficiency.

Graphite furnace atomization uses electric current to heat the graphite furnace atomizer to a high temperature of more than 2000℃, and the sample will be atomized in the furnace. The graphite furnace heating procedure is completed in 4 steps including drying, ashing, atomization, and decontamination. The temperature, temperature holding time, and heating method of each stage are determined according to the sample composition and analysis elements. The purpose of drying is to remove the solvent and avoid sample splashing during ashing and atomization. The function of ashing is to destroy volatile matrices and organic matter to reduce molecular absorption. At the atomization stage, the temperature is the highest, and the analytical elements evaporate and dissociate into ground-state atomic vapor. During atomization, the inert shielding gas stops to extend the residence time of ground-state atoms in the graphite tube, thereby improving the analytical sensitivity of the method. Decontamination is the removal of matrix residues left in the graphite tube at high temperatures, eliminating memory effects in preparation for the next assay.

Graphite furnace atomic absorption spectrometry (GFAAS) has poor precision and requires more operator skill. However, the advantages of this method are: the sample dosage is small, the atomization efficiency is almost 100%, and the absolute sensitivity of the analysis is high. It is an ideal method for the analysis of trace amounts of cadmium in soil.

Equipment and reagents

(1) **Equipment.** Atomic absorption spectrometer (AA800, PE company of the United States), cadmium hollow cathode lamp, oil-free air compressor.

(2) **Reagents.** High-purity argon, hydrochloric acid, nitric acid, hydrofluoric acid, and perchloric acid are all high-grade purity, cadmium powder (99.99%), ultrapure water and soil sample.

Procedures

(1) **Preparation of soil sample solution.** Accurately weigh $0.50 \sim 1.0$g soil sample in a 25mL polytetrafluoroethylene (PTFE) crucible, moisten with a little water, add 10mL HCl, and heat ($<450^{\circ}C$) on a hot plate for 2h. Then add 15mL HNO_3 and continue to heat until about 5mL of the dissolved analyte remains. Add 5mL HF and heat to remove the silicon compound. Finally, add 5mL $HClO_4$ to heat until the digestate is light yellow. Open the lid, and heat to near dryness. Remove the crucible from the hot plate and cool it down to room temperature. Add 1 : 5 HNO_3 1mL, dissolve the residue with a slight heat, transfer the solution into a 50mL volumetric flask and dilute to the mark.

(2) **Preparation of cadmium standard solution.** Cadmium stock solution: weigh 0.5000g of cadmium metal powder (spectrally pure), and dissolve in 25mL 1 : 5 HNO_3 (dissolved with heat). Cool in air, transfer the solution into a 500mL volumetric flask and dilute to the mark with deionized water. This solution contains 1.0mg/mL of cadmium.

Standard solutions of cadmium: pipette 10.0mL of the standard stock solution into a 100mL volumetric flask, dilute with water to the mark, and shake well for further use. Pipette 5.0mL of the diluted standard solution into another 100mL volumetric flask and dilute it with water to the mark to obtain a standard solution containing $5\mu g/mL$ of cadmium.

(3) **Instrument warm-up.** Turn on the argon switch, circulating water cooler ($18 \sim 22^{\circ}C$), graphite furnace power supply, and atomic absorption spectrometer power supply. Turn on the instrument workstation and the element lamp. Start the software and preheat the instrument for 40min.

(4) **Setting the operating parameters.** The source of radiation is a Cd hollow cathode lamp (228.8nm), the spectral bandwidth is set to 1.0nm, and the lamp current to 1mA. The argon flow rate is 0.2L/min, and the injection volume is $20\mu L$. The temperature programs for the graphite furnace are shown in Table 3-3 (for reference only).

Table 3-3　Temperature programs of the graphite furnace in determination Cd in soil

Step	Temperature/℃	Hold/s	Ramp/(℃/min)	Flow/(L/min)
Drying	100	30	10	0.2
Ashing	300	20	150	0.2
Atomization	900	30	0	0
Cleaning	2500	30	0	0.2

(5) Sample determination

① Working curve method. Pipette cadmium solution 0mL, 0.50mL, 1.00mL, 2.00mL, 3.00mL, 4.00mL in six 50mL volumetric flasks, respectively. Dilute all the flasks to the mark with 0.2% HNO_3 solution and shake well. This standard series contain cadmium $0\mu g/mL$, $0.05\mu g/mL$, $0.10\mu g/mL$, $0.20\mu g/mL$, $0.30\mu g/mL$, $0.40\mu g/mL$ respectively. Measure the absorbance sequentially. Under the same conditions, measure the absorbance of soil samples.

② Standard addition method. Pipette 5.00mL of the sample solution into each of the five 10mL volumetric flasks. To each of the flasks, add cadmium standard solution $(5.0\mu g/mL)$ 0mL, 0.50mL, 1.00mL, 1.50mL and 2.00mL respectively, dilute to the mark with 0.2% HNO_3 solution. Measure the absorbance sequentially.

Data processing

(1) Take the absorbance of each substance in the standard curve method as the ordinate and the concentration of each substance as the abscissa, obtain the standard curve and linear regression equation. Substitute the absorbance of the soil sample (minus the blank absorbance) into the regression equation and calculate the cadmium content in the soil.

(2) Plot the A-c standard curve with the concentration of the added standard solution as the abscissa and the measured absorbance as the ordinate. The epitaxial curve intersects the abscissa, and the distance between the origin and the intersection point is the concentration of diluted cadmium ions. Use the concentration to calculate the cadmium content in the soil.

Notes

(1) During oil sample digestion, the solution must be prevented from evaporating when removing $HClO_4$. When accidentally evaporated, Fe and Al salts may form insoluble oxides and contain cadmium, resulting in low results.

(2) The determination wavelength of cadmium is 228.8nm (in the ultraviolet region), which is easy to be interfered with light scattering and molecular absorption, especially at $220.0 \sim 270.0$nm, where NaCl has strong molecular absorption. It will overlap with the wavelength of cadmium at 228.8nm. In addition, the molecular absorption and light scattering of Ca and Mg are also very strong. These lead to an increased absorbance of cadmium. In order to eliminate matrix interference, an appropriate amount of matrix modifier can be added to the measurement system, such as adding 0.5g $La(NO)_3 \cdot 6H_2O$ to the standard series of solutions and samples, respectively. This method is suitable for the determination of cadmium content in cadmium-contaminated soil.

(3) The purity of perchloric acid has a great influence on the blank absorbance, which is directly related to the accuracy of the measurement results. Therefore, it is essential to correct the blank value in the whole process and minimize the amount of perchloric acid addition to reduce the blank value.

Questions

(1) How do different graphite furnace programs affect the measurement results?

(2) What is the effect of incomplete soil digestion on the measurement results?

Exp. 16　Determination of Selenium in Selenium Supplements by Graphite Furnace Atomic Absorption Spectrometry

Objectives

(1) To further master the operation skills of graphite furnace atomic absorption spectrometer.

(2) To master the function of matrix modifiers.

Principles

Selenium (Se) is an essential trace element for humans. It has been implicated with the protection of body tissues against oxidative stress, maintenance of defense against infection, and modulation of growth and development. Chronic Se deficiency may enhance susceptibility to viral infection, cancer, cardiovascular disease, thyroid dysfunction and various inflammatory conditions. People with selenium deficiency can use selenium supplement reagents to quantitatively and accurately supplement selenium according to human needs.

Selenium is a volatile element. In this experiment, low ashing temperature, matrix modifier and graphite furnace temperature programs are employed to enhance the analysis sensitivity and stability.

Equipment and reagents

(1) Equipment. TAS-990AFG graphite furnace atomic absorption spectrophotometer, Se hollow cathode lamp, volumetric flasks.

(2) Reagents. Concentrated HNO_3, 1000mg/L Se stock solution, $Ni(NO_3)_2 \cdot 6H_2O$, selenium dietary supplement tablets.

Procedures

(1) Preparation of standard solutions

① Weigh an appropriate amount of $Ni(NO_3)_2 \cdot 6H_2O$, dissolve it in deionized water and dilute to 50mL to obtain 5% Ni solution.

② Pipette an appropriate amount of Se stock solution (1000mg/L), and dilute to 500mL with deionized water to obtain 1mg/L Se stock solution.

③ Accurately pipette an appropriate amount of 1mg/L Se stock solution, add 0.5mL of concentrated HNO_3 and 2mL of 5% Ni solution to prepare five 100mL Se standard solutions with the concentrations of $0\mu g/L$, $5\mu g/L$, $10\mu g/L$, $20\mu g/L$ and $40\mu g/L$, respectively.

(2) Preparation of sample solution. Weigh one selenium dietary supplement tablet in a 250mL glass beaker, add 50mL of deionized water, and then add enough concentrated HNO_3 to make the acid concentration be 1%. Heat it at 95℃ for 1h and then cool it to room temperature. Transfer all the solution to a volumetric flask, and dilute to 100mL with deionized water.

(3) Setting operating parameters. The lamp current is 9.0mA, and the spectral bandwidth is 1.0nm. The analytical line is set at 196.0nm, and the sample injection volume is 10μL. The temperature programs of the graphite furnace are shown in Table 3-4.

Table 3-4　Temperature programs of the graphite furnace for determining selenium

Step	Temperature/℃	Hold/s	Ramp/(℃/min)	Flow/(L/min)
Drying 1	90	10	5	3.0
Drying 2	105	20	3	3.0
Drying 3	300	10	50	3.0
Ashing	1100	5	50	3.0
Atomization	2100	4	1400	0
Cleaning	2300	4	500	3.0

(4) Analysis of unknown sample

① Turn on the TAS-990AFG graphite furnace atomic absorption spectrophotometer and computer. Set up the operating parameters of the optimized method.

② Determine the absorbance of the selenium-containing standard solutions sequentially.

③ According to the linear range of the standard curve, appropriately dilute the unknown sample if necessary. Determine the absorbance of the sample and the blank solution and repeat the assay three times.

Data processing

(1) Plot the standard curve and calculate the concentration of Se in the diluted unknown sample.

(2) Calculate the Se content in the Se dietary supplement and compare the result with that of the manufacturer.

Questions

(1) What is the function of nickel nitrate solution in this experiment?

(2) Design an experimental protocol for measuring the selenium content in real samples using the standard addition method.

Exp. 17　Determination of Cd and Cr in Wastewater by ICP-AES

Objectives

(1) To master the components and basic operation skills of ICP-AES.

(2) To master the principle of ICP light source.

(3) To master the preparation method of multi-element standard solutions.

Principles

Atomic emission spectroscopy is an analytical method that determines the composition and content of substances based on the characteristic wavelengths and the emission intensity emitted by gaseous atoms or ions of elements after excitation. In the excitation light source, the measured substance is vaporized, dissociated, ionized, excited, and radiation is thus generated. Currently, the commonly used light sources are arc, spark and inductively coupled plasma (ICP). Among them, ICP has the advantages of strong excitation, good stability, small matrix effect and low detection limit. Meanwhile, ICP has no self-absorbance and has a wide linear range.

Inductively coupled plasma-atomic emission spectroscopy (ICP-AES), also known as ICP-OES (optical emission spectroscopy), is a type of emission spectroscopy that is often used to simultaneously detect multi-elements in the unknown sample. It excites the sample in a plasma source and causes the measuring element to emit characteristic wavelength radiation. Qualitative analysis is done according to its wavelength position, and quantitative analysis is performed by measuring the emission intensity of the elements. Since the emission intensity is directly proportional to the concentration of the element, the standard curve method, standard addition method and internal standard method can be used for quantitative analysis in real sample analysis.

Cadmium and chromium in wastewater are the first class of pollutants in water and are extremely harmful to humans. In this experiment, the standard curve method is used to simultaneously determine the content of Cd and Cr in wastewater after the sample is pretreated.

Equipment and reagents

(1) **Equipment.** Agilent 5100 ICP-OES spectrometer, beaker, volumetric flask, electric stove, sonicator.

(2) **Reagents.** 100mg/L cadmium standard stock solution, 100mg/L chromium standard stock solution, nitric acid (HNO_3), perchloric acid ($HClO_4$), industrial wastewater.

Procedures

(1) **Setting the operating parameters.** The measuring wavelength for Cd and Cr are 226.502 and 267.716nm, respectively. The incident power is 1kW. The flow rate of argon

cooling gas, auxiliary gas and carrier gas are 12-14L/min, 0.5-0.8L/min, and 1.0L/min, respectively.

(2) Preparation of standard solutions. Prepare 0.1mg/L, 0.5mg/L, 1.0mg/L, 5.0mg/L, 10mg/L series of bi-element standard solutions of cadmium and chromium.

(3) Pretreatment of wastewater. Pipette 50mL of water sample into a 100mL beaker, and heat it to about 10mL on an electric stove. Add 10mL of mixed acid (HNO_3 : $HClO_4$ = 4 : 1), vaporize it until near dryness, then add 5mL HNO_3 (1 : 1) and a small amount of deionized water to about 25mL. After cooling down to room temperature, ultra-sonicate it for 0.5h to dissolve the residue. Transfer the filtrate into a 50 mL volumetric flask after filtration, dilute to the mark, and shake well.

(4) Determination of sample

① Follow the operation steps to power the instrument on and ignite the ICP torch.

② Select Cd and Cr elements in the periodic table, the measuring wavelength and the optimized working conditions.

③ Spray different concentrations of mixed standard solutions sequentially, measure their emission intensities, and the instrument automatically plots a standard curve.

④ Spray into the industrial wastewater test solution and collect data. The instrument automatically calculates the results.

⑤ According to the shutting down procedure, exit the analysis program. Go to the main menu, turn off the peristaltic pump, and air circuit. Turn off the ICP power supply and computer, and finally turn off the cooling water.

Notes

(1) Once the detection is over, the sample injection system should be sprayed with deionized water for 3min, and then power off to avoid deposition of samples on the nebulizer mouth and quartz torch tube nozzle.

(2) The plasma emits strong ultraviolet light and is easy to damage the eyes, so the ICP torch should be observed through the colored glass protective window.

Questions

(1) Why can ICP improve the detection sensitivity and accuracy of AES?

(2) Briefly describe the similarities and differences between ICP-AES and flame atomic absorption spectroscopy.

Exp. 18　Determination of Impurity Elements in Pure Zinc Samples by ICP-AES

Objectives

(1) To further grasp the basic principles of ICP-AES.

(2) To further master the analytical method for the simultaneous determination of multiple elements using ICP-AES.

Principles

Pure zinc samples usually contain Pb, Cd, Fe, Cu, Sn, Al, As, Sb and other impurity elements, and the impurity content is a key indicator in determining the product quality. Determining the content of these elements by spectrophotometry or flame atomic absorbance is cumbersome and time-consuming. Inductively coupled plasma-atomic emission spectroscopy (ICP-AES) can perform multi-element analysis at the same time, which has the advantages of fast analysis speed, high sensitivity, good stability, wide linear range and small matrix interference. Using ICP-AES to determine the impurity elements in pure zinc samples can not only greatly improve the analysis efficiency, but also make the analysis results more accurate and feasible.

Equipment and reagents

(1) **Equipment.** Agilent 5100 ICP-OES spectrometer, beaker, electric hot plate, volumetric flasks.

(2) **Reagents.** 1.000mg/mL multi-element standard stock solution, concentrated hydrochloric acid (AR), ultrapure water, pure zinc sample.

Procedures

(1) **Preparation of pure zinc sample solution.** Accurately weigh about 0.5g of pure zinc sample to a 100mL beaker. Add 10mL of 1 : 1 hydrochloric acid to the beaker, cover it with a watch glass, and heat it on an electric hot plate. After the zinc particles are dissolved, vaporize the solution to near dryness, and rinse the inner wall of the watch glass and beaker with a washing bottle. After cooling to room temperature, transfer all the solution to a 25mL volumetric flask, dilute it with 5% HCl ultrapure aqueous solution, and shake well.

(2) **Preparation of multi-element standard solutions.** Prepare series multi-element standard solutions with the concentration of 0.1μg/mL, 0.5μg/mL, 1.0μg/mL, 5.0μg/mL, 10.0μg/mL, respectively. Dilute all the solutions with 5% HCl ultrapure aqueous solution.

(3) **Setting operating parameters.** The high-frequency power is 1150W. The flow rate of argon cooling gas, auxiliary gas and carrier gas are 12L/min, 0.5L/min, 1.0L/min, respectively. The peristaltic pump speed is 50r/min.

(4) **Sample analysis**

① Follow the basic operation steps of the 5100 ICP-OES spectrometer to complete the preparation, start and ignite the plasma.

② Determine the emission intensity of the multi-element standard solutions, and the instrument will automatically plot the standard curves of each element.

③ Spray in the prepared pure zinc sample solution, the instrument automatically gives the measurement results. According to the measurement results, calculate the mass fraction

(%) of the impurity element.

$$w_x = \frac{\rho V \times 10^{-6}}{m} \times 100\%$$

Where ρ is the mass concentration of impurity elements in pure zinc samples, the unit is $\mu g/mL$; V represents the volume of the solution in mL; m stands for the mass of the sample in grams.

④ Once the entire run is complete, allow 5% HCl ultrapure aqueous solution to be pumped through the system for 5min and ultrapure water for another 5min, and then extinguish the plasma. Turn off the cooling water 5min later, and turn the argon off after the CID detector rises to room temperature. Finally, turn off the exhaust.

Notes

(1) During the sample dissolution process, wait for the solution to cool down before transferring it to the volumetric flask, to avoid errors in the volume.

(2) If the salt content of the sample is high, the atomization should be observed at any time to prevent the nebulizer mouth from being blocked. If the sample is not completely dissolved, it is strictly forbidden to be measured on the machine.

Questions

(1) Briefly describe the principle and process of ICP generation.

(2) How to select the working parameters when performing a multi-element simultaneous analysis?

Exp. 19   Simultaneous Determination of Trace Multi-element in Ham by Microwave Digestion-ICP-AES

Objectives

(1) To master the sample pretreatment method by microwave digestion.

(2) To be familiar with the operation skills of microwave digestion instruments.

(3) To master the qualitative and quantitative analysis method for simultaneous detection of multiple elements in ICP-AES.

Principles

Jinhua ham is a famous specialty of Zhejiang Province. Ham meat is ruddy in color and rich in aroma and is well-known at home and abroad for its color, fragrance, taste and shape. Ham is rich in nutrition. It not only contains a lot of protein and fat but also contains a variety of essential elements such as potassium, calcium, iron, etc. The study on the content of trace elements in ham plays an important role in the evaluation of the quality and nutritional value of ham.

Microwave digestion is a sample pretreatment technology, which digests samples in a high-pressure closed container. This pretreatment technology endows the measurement elements to be measured not contaminated, volatile elements are not easy to lose. The sample digestion is complete, fast, and reagent-saving as well. Microwave digestion combined with ICP-AES can sensitively realize the rapid determination of trace and ultra-trace elements in environmental, high-purity materials, geology, nuclear science, biology, medicine, chemometrics, agriculture and food research.

In this experiment, microwave digestion is first used to pretreat the ham sample, and then the content of potassium (K), calcium (Ca), sodium (Na) and magnesium (Mg) in ham is quantitatively determined by the standard curve method using inductively coupled plasma-atomic emission spectroscopy (ICP-AES).

Equipment and reagents

(1) **Equipment.** Agilent 5100 ICP-OES spectrometer, ETHOS microwave digester.

(2) **Reagents.** Dry ham, nitric acid, 1000mg/L K, Ca, Na, Mg standard stock solution.

Procedures

(1) Sample pretreatment

① Grind the ham meat in a meat grinder for further use.

② Accurately weigh the meat sample of about 0.2000g into the polytetrafluoroethylene (PTFE) digestion tank (3-5 samples in parallel), add a certain amount of nitric acid, and tighten the digestion tank. Set the parameters in Table 3-5 and start sample digestion. After the program is over, automatically cool the sample and take out the digestion tank. Put the tank on the heater and drive away the acid. After cooling down to room temperature, transfer the sample solution into a 25mL colorimetric tube and dilute it with deionized water. Prepare 3 blank samples as control.

Table 3-5 Microwave digestion program for ham

Steps	Temperature variation /℃	Time /s	Power /W
1	Room temperature~120	300	500
2	120	300	500
3	120~195	600	800
4	195	600	800
5	Cooling	600	0

(2) **Working conditions for elemental determination.** Follow the basic operating procedures of the 5100 ICP-OES spectrometer to complete the machine startup and plasma igniting. Some of the working parameters are as follows: the power is 1.15kW, the atomization gas flow is 0.75L/min, the auxiliary gas flow is 1.5L/min, and the plasma gas flow is

15L/min. The observation position is automatically optimized, and the observation direction is axial. The analytical lines for each element are shown in Table 3-6.

Table 3-6　Analytical lines for each element

Elements detected	Wavelength/nm
K	766.491
Ca	422.673
Na	589.592
Mg	279.533

　(3) **Plotting the standard curve.** Prepare the series multi-standard solution from stock solution with 5% nitric acid in gradients of 0.00μg/mL, 5.00μg/mL, 10.00μg/mL, 20.00μg/mL, 50.00μg/mL, 100.0μg/mL. Under the selected instrument conditions, spray different concentrations of multi-standard solutions into the instrument sequentially, and collect the data. The standard curve of each element is automatically obtained by the instrument.

　(4) **Sample determination.** Spray in the prepared ham sample solution, collect data, and the instrument automatically gives the measurement results of each element.

　(5) **Shutting down.** After the analysis procedure is completed, rinse the instrument with 5% HCl and ultra-pure water for 5min, respectively. Then extinguish the plasma. Turn off the cooling water after 5min, and turn off the argon gas after the detector temperature rises to room temperature. Finally, turn off the exhaust.

Data processing

　Calculate the content of each element in ham (μg/g) according to the measurement results.

Questions

　(1) What are the advantages of using ICP-AES for multi-element analysis?

　(2) Describe the methods for sample pretreatment.

Exp. 20　Determination of Metal Elements in Purified Drinking Water by Inductively Coupled Plasma Mass Spectrometry

Objectives

　(1) To understand the basic principles of how ICP-MS works.

　(2) To understand the instrument structure and operation method of ICP-MS.

Principles

Inductively coupled plasma mass spectrometry (ICP-MS) is an inorganic multi-element analysis technique that uses inductively coupled plasma as an ion source and mass spectrometry as the detector. It is an analytical technique that combines ICP technology and mass spectrometry, which can simultaneously determine dozens of trace inorganic elements and can perform isotopic analysis, single-element and multi-element analysis and morphological analysis of metal elements in organic matter.

The working principle of ICP-MS is that the sample is first brought into the atomization system by the carrier gas (argon) for atomization. Then it enters the axial channel of the plasma in the form of an aerosol and is fully vaporized, atomized and ionized in high-temperature and inert gas. And then the generated ions enter the vacuum system through the sampling cone and skimmer cone, which are focused by the ion lens, and separated by the quadrupole mass spectrometer according to the mass-to-charge ratio. Qualitative and quantitative analyses are performed according to the position of the mass spectrum and the concentration of the elements, respectively.

Purified drinking water is water that contains no additives and is safe to drink directly. Excessive metal elements in drinking pure water directly affect the health of human beings. Monitoring the content of metal elements in drinking pure water and ensuring the quality of drinking water are of great significance for maintaining the health of consumers.

The commonly used methods for detecting metal elements in drinking water include atomic absorption spectroscopy, atomic fluorescence method, spectrophotometry, ICP-AES, ICP-MS, etc. Among them, ICP-MS is the best choice for the simultaneous determination of multi-metal elements in water quality due to its advantages of high sensitivity, low detection limit, wide linear range and simultaneous determination of multiple elements.

Equipment and reagents

(1) **Equipment.** iCAP Q inductively coupled plasma mass spectrometer, Milli-Q ultra-pure water system.

(2) **Reagents.** Purified drinking water comes from commercially available products, nitric acid, Tl, Ba, Sb, Sn, Cd, Se, Cu, Ni, Co, Mn, V, Ti, Cr, Mo, B, Al, Zn, Fe, Mg, Ca, As, Sr, Hg, Pb multi-element standard solutions (100mg/L National Reference Standards Center).

The internal standard element is Sc, Ge, Rh, Bi (1000μg/L, National Reference Standards Center).

Procedures

(1) **Sample preparation.** Pipette 10mL of purified water, add 1 drop of nitric acid and mix well for further use.

(2) **Setting ICP-MS operating parameters.** The radiation frequency (RF) power is 1500W. The plasma gas flow rate is 15L/min, the carrier gas flow rate is 0.80L/min, and

the auxiliary gas flow rate is 0.40L/min. The atomization chamber temperature is 2℃, and the nebulizer is concentric. The sampling cone and skimmer cone is nickel/platinum cone, and the acquisition mode is the peak detect mode.

(3) **Preparation of standard solution and selection of internal standard.** Take a certain amount of multi-element standard stock solution, and sequentially dilute it with 2% HNO_3 to obtain the standard solutions in the range of 2-50μg/L. The selection of internal standard elements should be close to the mass of the measuring element. Sc, Ge, Rh, and Bi are the internal standard elements. The internal standard stock solution has a concentration of 20μg/L diluted with 2% HNO_3. The internal standards selected according to the different elements are:

① B, Mg, Al, Ca, and Fe use Sc as the internal standard.

② Ti, Zn, V, Cr, Mn, Co, Ni, and Cu use Ge as the internal standard.

③ As, Se, Sr, and Mo use Rh as the internal standard.

④ Cd, Sn, Sb, Ba, Hg, Tl and Pb use Bi as the internal standard.

(4) **Sample determination.** After checking the vacuum level, argon pressure, circulating cooling water and other parameters, ignite the plasma. After stabilization, optimize the parameters of the instrument. Through the three-way valve, inject the internal standard and the sample in equal volume. Edit and tune in sequences for sample analysis.

Data processing

Calculate the amount of trace elements in the water sample.

Notes

(1) The utensils used are soaked with 10% HNO_3 for 24h, then rinsed with ultrapure water before use.

(2) This method can detect 24 elements such as arsenic, copper, lead, iron, manganese, zinc, cadmium, aluminum, mercury, barium, boron, molybdenum, sodium, selenium, silver, nickel, thallium, beryllium and antimony. If there are no sufficient standards, fewer trace elements can also be selected for measurement.

Questions

(1) What are the commonly used determination methods of metal elements? Compare their pros and cons.

(2) What are the mass spectrometry interferences in the determination of trace elements in water by inductive plasma mass spectrometry? How to eliminate them?

Exp. 21　Determination of pH of an Unknown Sample Using Potentiometry

Objectives

(1) To Learn the principles of using direct potentiometry to determine the pH of a solution.

(2) To learn using the pHS-3C pH meter.

Principles

Potentiometry is an analytical method that directly calculates the activity (or concentration) of the measured substance from the relationship of the Nernst equation based on the potential value of the indicator electrode that makes up the electrochemical cell. It is commonly used for the accurate determination of the pH of an aqueous solution. IUPAC recommends direct comparison with standard buffer solutions for the assay. For the measurement, the glass electrode and the reference electrode are combined into the following batteries:

Hg, Hg_2Cl_2 | saturated KCl solution ‖ analyte | glass membrane | inner reference solution | AgCl, Ag

The potential of the cell can be defined as:

$$E_{cell} = E_{glass} - E_{SCE} + E_j$$

Where E_{SCE} is the potential of the reference electrode, E_{glass} represents the potential of the glass electrode, which can reflect the activity (concentration) information of the substance to be measured. E_j is the junction potential through the salt bridge. Under certain conditions, E_{SCE} and E_j are constants.

E_{glass} can be described as:

$$E_{glass} = k - 0.059pH$$

The cell's potential is thus abbreviated as

$$E_{cell} = K - 0.059pH$$

If K is known, the pH of the unknown solution can be determined from the measured E value. However, the K value is not easy to obtain. We can compare the cell's potential of the test solution and the standard solution (known pH) to determine the pH of the unknown. This method is called a direct comparison method or a two-point calibration method. 0.059V/pH (or 59mV/pH) is called the response slope of the pH glass electrode (25℃). The ideal pH glass electrode has a slope near 59mV/pH at 25℃. However, due to differences in production processes, the slope of each pH glass electrode may be different, and it must be determined by experimental methods.

Equipment and reagents

(1) **Equipment.** pHS-3C acidity meter, pH combination electrode (Raymond E-201-C), magnetic stirrer, and stirring magnet.

(2) **Reagents.** Potassium hydrogen phthalate standard buffer solution (pH=4.00), potassium dihydrogen phosphate and disodium hydrogen phosphate standard buffer solution (pH=6.86), borax standard buffer solution (pH=9.18), pH unknown solution. Standard buffer reagents are all commercially bagged and the standard solutions can be prepared as required.

Procedures

(1) **Calibration.** For precision pH meters, in addition to the "Position" and "Temperature Compensation" adjustments, there is an electrode "Slope" adjustment that can be calibrated with two standard buffers. Normally, a "Position" calibration is performed with a

buffer solution of pH=6. 86, and then a "Slope" correction is performed using a pH=4. 00 (acidic) or pH=9. 18 (alkaline) buffer solution depending on the pH of the unknown solution.

① Take the combination electrode out of its storage solution, rinse it with deionized water, and dry it with a filter paper. Immerse the dried electrode into a standard solution with pH=6. 86 and use a thermometer to measure the temperature of the solution. Set the temperature on the equipment. Press the "Position" button, and then press the "Confirm" button. The pH meter will automatically recognize the current pH value of the solution.

② Rinse the electrode with deionized water, and dry it with a filter paper. Immerse the electrode into the second standard solution (near the expected sample, either a pH of 4. 00 or 9. 18), set the temperature, and press the "Slope" button to start the calibration. Again, the equipment will automatically identify the current pH value of the test solution.

(2) Determination of pH of the unknown sample. The calibrated machine can measure the pH of unknown solutions and no more pressing the "Position" and "Slope" keys. When the temperature of the test solution is the same as that of the standard solution, rinse the electrode with deionized water, and dry it with a filter paper. Insert the electrode in the unknown solution, stir the liquid, allow the reading to stabilize, and read the pH.

(3) Measuring the response slope. Switch the pH function button to the mV mode, insert the electrode into the standard buffer solution at pH=4. 00, read the mV value of the solution on the display, and then measure the mV value of the pH buffer solution at pH= 6. 86 and pH=9. 18, respectively.

Data processing

(1) Record the pH value of the unknown solution.

(2) Plot a graph with E against pH, and the slope of the straight line is the slope response of the glass electrode. If it deviates too much from 59mV/pH (25℃), the electrode cannot be used.

Notes

Glass electrodes are fragile, therefore do not allow the electrode to touch the bottom of the beaker or the stirring bar to hit it.

Questions

(1) When measuring pH, why should we use a standard buffer solution with a pH close to that of the test solution?

(2) Can we use a millivolt meter to measure the solution's pH?

Exp. 22 Determination of the Total Fluorine Content in Toothpaste Using Fluoride Ion Selective Electrode

Objectives

(1) To master the determination method of fluoride ion content in toothpaste.

(2) To master the composition and function of the total ionic strength adjustment buffer.

Principles

Fluorine is the most active non-metallic element. It is also one of the essential trace elements of the human body. An appropriate amount of fluoride is beneficial to the human body, too low intake will produce dental caries, but the long-term intake of excess normal needs will lead to endemic fluorosis. Potentiometry with fluoride ion selective electrodes is one of the commonly used methods for measuring fluoride ions. This method has the advantages of simple and firm electrode structure, high sensitivity, and fast response speed. It is widely used due to free color interference, high precision, and ease of operation.

Fluoride ion selective electrode (FISE) is a kind of crystal homogeneous film electrode, made of LaF_3 single crystal, which can specifically identify fluoride ions and is an indicator electrode for measuring the activity of fluoride ions in solution by potentiometry. The potential of fluoride electrode (E_F) is

$$E_F = K - \frac{RT}{nF} \ln a_{F^-}$$

When the fluoride ion selective electrode and the reference electrode form a cell, the signal of the instrument, namely the potential of the cell (E_{cell}) is linearly corresponding to $\ln a_{F^-}$. If an appropriate amount of ionic strength adjuster is added to the test solution to keep the ionic strength unchanged, the activity coefficient is constant. Thus, the ion activity can be replaced by concentration, and the E_{cell} has a linear relationship with lg [F^-]. According to the standard curve of E_{cell} to lg [F^-] and E_{cell} of the sample, we can obtain the fluoride ion concentration in the sample. In potentiometry, the total ionic strength of the solution is usually controlled by adding a total ionic strength adjustment buffer (TISAB).

Equipment and reagents

(1) **Equipment.** Fluorine ion selective electrode, saturated calomel electrode, acidity meter, electromagnetic stirrer.

(2) **Reagents**

① 1.0×10^{-3} mol/L F^- standard stock solution. ②TISAB: Weigh 58g of NaCl, 12g of sodium citrate $Na_3C_6H_5O_7 \cdot 2H_2O$, take 57mL of ice vinegar, and dissolve them in 500mL deionized water. Slowly add 6mol/L NaOH solution, and adjust the pH to 5.5-6.5. After cooling, transfer the solution to a 1000mL volumetric flask, dilute to the mark with water, shake well and store in a polyethylene bottle. ③Fluoride toothpaste.

Procedures

(1) **Sample preparation.** Accurately weigh 0.9-1.4g of fluoride toothpaste and put it in a small plastic beaker. Add an appropriate amount of deionized water and stir thoroughly, and then ultrasonic the solution for about 20 minutes. Transfer all the solution to a 100mL volumetric flask, add 10 mL of TISAB solution, add water to the mark and mix well for further use.

(2) Instrument startup. Preheat the pH meter for 20min, and connect the fluorine ion-selective electrode and saturated calomel electrode to the pH meter, respectively. Immerse the two electrodes into the deionized water, start the stirrer and clean the electrode repeatedly to the blank potential (about $-300mV$).

(3) Preparation of a calibration curve. Accurately pipette 0.5mL, 1.00mL, 2.50mL, 5.00mL, 10.00mL of standard F^- solution ($1.0 \times 10^{-3} mol/L$) into five 100mL volumetric flasks, respectively. Add 10mL of TISAB solution into each of the flasks, and dilute to the mark with deionized water. Transfer the series of standard solutions from low concentration to high concentration into a dry plastic cup, put in the stirring rod, insert the electrode into the test solution, and start the stirrer for 5-8min. Stop stirring, and read the equilibrium potential.

(4) Sample determination. Transfer the treated toothpaste sample solution to a dry plastic cup and measure the E value.

Data processing

(1) Make a standard curve by plotting E *vs.* lg $[F^-]$ to obtain a linear regression equation.

(2) Substitute the potential value of the toothpaste sample into the linear regression equation and calculate $[F^-]$, the fluorine content in the toothpaste sample.

Questions

(1) What is the purpose of adding TISAB in this experiment?

(2) Why wash the fluorine electrode to a certain potential?

Exp. 23   Measuring Selectivity Coefficient of Chloride Ion Electrode to Bromine Ion

Objectives

(1) To understand the principle and method of determination of selectivity coefficient of an ion selectivity electrode.

(2) To master the experimental technique of measuring the selectivity coefficient of ion selectivity electrode by mixed solution method.

Principles

An ion-selective electrode is an electrode that responds selectively to a specific ion. But this selectivity is not totally specific, and the potential of other coexisting ions in the solution to the electrode may also produce some "contribution". The difference in the response of measuring ions and other ions of the ion-selective electrodes can be quantitatively characterized by the selectivity coefficient. If the measured ion is i, j is the interference ion, n and m are the number of charges of them, respectively. then $K_{i,j}$ is the selectivity coefficient of the electrode. The contribution of coexisting ions to potential can be described using Nicolsky's equation:

$$E = K \pm \frac{2.303RT}{nF} \lg(a_i + K_{i,j} a_j^{n/m})$$

As can be seen from the above formula, the smaller the selectivity coefficient, the less interference by j ion. $K_{i,j}$ can be determined by the separate or mixed solution method. Herein, we use the mixed solution method to determine $K_{i,j}$. Specifically, a series of standard solutions containing fixed activity of interference ions (j) and different activities of measured ions (i) are prepared, and the corresponding potential values E are measured separately to plot an E-$\lg a_i$ curve.

When a_i is greater than a_j, the electrode to i ion is a Nernstian response, and the influence of interfering ions is negligible. If the i, j ions are both monovalent anions (e. g. , in this experiment), the Nernst equation for the straight line of the standard curve is:

$$E_1 = K_1 - \frac{2.303RT}{nF} \lg a_i$$

When a_i is less than a_j, the standard curve reaches a platform. The response of the electrode to the i ion is negligible, and the potential value is completely determined by the j ion.

$$E_2 = K_2 - \frac{2.303RT}{nF} \lg(K_{i,j} a_j^{n/m})$$

Assuming that $K_1 = K_2$ and the two slopes are the same, then at the intersection of the two lines $E_1 = E_2$. $K_{i,j}$ can be expressed as:

$$K_{i,j} = a_i / a_j^{n/m}$$

Therefore, the value $K_{i,j}$ can be obtained by the method of the fixed interference method. In this experiment, the selectivity coefficients of the chloride ion selective electrode were determined by using Br^- as the interfering ion.

$$K_{Cl^-, Br^-} = a_{Cl^-} / a_{Br^-}$$

The measurement system of this experiment consists of a chloride-ion selective electrode, a reference electrode and a test solution. The sensitive membrane of the chloride-ion selective electrode is made of Ag_2S-AgCl powder mixed tablet. It is a fully solid electrode without an internal reference solution, and the charge is conducted by Ag^+, which has the fewest number of charges and the smallest radius in the film. Since the Cl^- in the saturated calomel electrode (SCE) can diffuse into the solution through a porous ceramic core, affecting the determination, therefore, a double salt bridge saturated calomel electrode should be used.

Equipment and reagents

(1) **Equipment.** pHS-3C pH meter, magnetic stirrer, chloride ion selective electrode, Type 217 double salt bridge SCE.

(2) Reagents

① 0.1000mol/L NaCl standard solution: Accurately weigh 1.464g of NaCl (AR, dried at 110℃) into a 50mL beaker, and dissolve it in deionized water. Transfer the solution into a 250mL volumetric flask, add water to the mark and mix well. ②0.1000mol/L NaBr standard solution: Accurately weigh 2.573g NaBr (AR) in a 50mL beaker, and add deionized water to dissolve it. Transfer the solution to a 250mL volumetric flask, dilute with water to the mark and mix well. ③ Using 1.0mol/L KNO_3 as the ionic strength adjustor and make the pH=2.5with HNO_3.

Procedures

(1) Machine startup

① Turn on the power switch of the pHS-3C pH meter and press the pH/mV key to the "mV" mode.

② Check whether the double salt bridge saturated calomel electrode is full of KCl solution, if not, add saturated KCl solution and exclude air bubbles.

③ Connect the chloride-ion selective electrode and the calomel electrode to the pHS-3C pH meter. Immerse the electrode into the deionized water, start the stirrer, and wash the electrode to a blank potential.

(2) Test solution preparation.
Accurately pipette the appropriate amount of chloride ion standard solution in a 50mL volumetric flask to prepare 1.00×10^{-4} mol/L, 1.00×10^{-3} mol/L, 5.00×10^{-3} mol/L, 1.00×10^{-2} mol/L, 5.00×10^{-2} mol/L and 1.00×10^{-1} mol/L NaCl series of standard solutions, add 5.00mL 1.00×10^{-2} mol/L Br^- standard solution and 15 mL 1.0mol/L KNO_3 solution into each flask. Dilute the solution to the mark with deionized water and shake well.

(3) Potential determination.
Transfer the series of test solutions from low concentration to high concentration into the beaker, respectively. Insert the electrode into the above solution, start the stirrer, and read the equilibrium potential of each solution after stopping stirring.

Data processing

Plot a curve with the measured potential E against $\lg c_{Cl^-}$, extend the two straight sections of the curve, and obtain the value of c_{Cl^-} from the intersection point. The selectivity coefficient of the chloride ion selective electrode for bromine ions is calculated according to the following formula.

$$K_{Cl^-,Br^-} = c_{Cl^-}/c_{Br^-}$$

Questions

(1) What are the performance parameters to evaluate an ion-selective electrode?

(2) Why should the double salt bridge saturated calomel electrode be used as the reference electrode in this experiment?

(3) Can the potential value of the chloride response be corrected by the selectivity coefficient? Why?

Exp. 24   Determination of the Acetylsalicylic Acid in Aspirin Tablet by Potentiometric Titration

Objectives

(1) To grasp the basic principles of potentiometric titration.

(2) To be familiar with the method of determining the endpoint of potentiometric titration.

Principles

Potentiometric titration is an analytical method that uses an instrument instead of the human eye to determine the end point of the titration by using a sudden jump in the indicating electrode potential. When carrying out the potentiometric titration, select the appropriate indicator electrode and the reference electrode to start an electrochemical cell with the tested solution. With the addition of titrant, the concentration of measured ions changes constantly due to chemical reactions, so the potential of the indicator electrode changes accordingly. Near the end point of the titration, the concentration of the measured ion is abruptly changed, causing a sudden jump in the electrode potential. Therefore, the titration end point can be determined based on the jump in electrode potential, without using chemical indicators to determine the end point.

Theoretically, any acid-base titration can be conducted by potentiometric titration. Two electrodes, after calibration, are immersed in a solution of the analyte. One is the pH indicator electrode and the other is the reference electrode. pH compound electrodes can also be used.

When conducting an acid-base titration using a pH meter, it is necessary to record the volume of titrant (V) and the corresponding pH value. By plotting the pH-V, $\Delta pH/\Delta V$-V or $\Delta^2 pH/\Delta V^2$-V graph to signal the end point and find out the concentration of the unknown samples.

In this experiment, aspirin tablets are dissolved in 95% ethanol solution, and the potentiometric titration is performed with NaOH standard solution. The volume V and pH of NaOH solution consumed at different times are recorded, and the content of acetylsalicylic acid in the sample is determined by the above-mentioned graphics.

Equipment and reagents

(1) **Equipment.** pH meter (pHS-3C), pH compound electrode, and magnetic stirrer with a Teflon-coated stirring bar.

(2) **Reagents.** Potassium hydrogen phthalate standard buffer solution (0.05mol/L, pH = 4.00), KH_2PO_4 and Na_2HPO_4 standard buffer solution (pH=6.86), 95% ethanol, deion-

ized water, NaOH solution (0.1mol/L), potassium hydrogen phthalate (KHP, AR).

Procedures

(1) Standardization of the pH meter. Standardize the pH meter using standard buffer solutions of pH = 4.00 and pH = 6.86 respectively. Refer to the calibration procedure to step (1) of Experiment 21.

(2) Standardization of NaOH solution. Using KHP as the primary standard, determine the exact concentration of NaOH standard solution.

(3) Potentiometry titration

① Grind 10 tablets of aspirin, and weigh accurately 0.6g of them. Put it into a 150mL beaker and dissolve it with 15mL of 95% ethanol.

② Add about 1.0mL of standard NaOH solution each time at the beginning of the titration till the pH value reaches 4.5, then followed by additional 0.2mL of standard NaOH solution till the pH value reaches 5.0. Reduce the amount of addition to one or half drop (0.02~0.04mL) near the stoichiometric point, when the pH changes sharply. Continue titrating to the end point until the pH change is less than 0.10.

Data processing

(1) Record the volume of NaOH added and the pH value of the solution at different times. Fill them in Table 3-7.

Table 3-7　The volume of NaOH and the pH value at different times

V_{NaOH}/mL	pH	ΔpH	ΔV	$\Delta pH/\Delta V$	$\overline{V}$	$\Delta^2 pH/\Delta V^2$

(2) Draw the pH-V titration curve (X-axis: volume of titrant added; Y-axis: pH of the sample).

(3) Calculate the content of the acetylsalicylic acid in the aspirin tablet ($M_{aspirin}$ = 180.2g/mol, $V_{NaOH,ep}$ is the volume of NaOH consumed at end point).

$$\omega = \frac{c_{NaOH} \times V_{NaOH,ep} \times M_{aspirin}}{m_{aspirin}} \times 100\%$$

Questions

(1) How to determine the end point according to $\Delta pH\text{-}V$, $\Delta pH/\Delta V\text{-}V$ and $\Delta^2 pH/\Delta V^2\text{-}V$ curves?

(2) Discuss the error in the experiment.

Exp. 25　Determination the Concentration of Weak Acids and Its Dissociation Constants by Automatic Potentiometric Titration

Objectives

(1) To understand the construction of the automatic potential titrator.

(2) To use an automatic potential titrator.

(3) To learn measuring the concentration and dissociation constant of acetic acid solution by automatic potentiometric titration.

Principles

An automatic potentiometric titrator consists of a syringe-like aspirator connected to the burette, a stirrer, two electrodes, and a titration vessel. The aspirator is used to hold the titrant solution or supplement the titrant solution, and the titration cup is used to contain the solution to be measured. In this experiment, the acetic acid solution is titrated against NaOH standard solution, the glass electrode is the indicator electrode, and the saturated calomel electrode (SCE) is the reference electrode. At 25℃, the potential of the cell is:

$$E=K+0.05916pH$$

Therefore, the pH of the solution has a linear relationship with the cell potential, and the change of E value can reflect the change in the pH of the solution. According to the titration curve, the end point can be found. The volume at the midpoint of the titration curve is the stoichiometric point.

Acetic acid (denoted by HAc) is a weak electrolyte. When $[Ac^-] = [HAc]$, $K_a = [H^+]$, the dissociation constant K_a of acetic acid is equal to the concentration of hydrogen ions in solution. If the volume of NaOH consumed at the end of the titration is V_{ep}, the corresponding hydrogen ion concentration is equal to K_a when the volume of NaOH consumed is 1/2 of V_{ep}.

Equipment and reagents

(1) **Equipment.** 888Titrando automatic potentiometric titrator, pH glass electrode, saturated calomel electrode, titration vessel, 25mL pipette.

(2) **Reagents.** 0.1mol/L NaOH solution, 0.05mol/L CH_3COOH solution, potassium hydrogen phthalate (KHP, AR).

Procedures

(1) **Machine startup.** Switch on the power supply of the instrument and preheat it. Install the instrument according to the operation instructions.

(2) **Standardization of NaOH solution.** Accurately weigh KHP 2.5g into a beaker, dissolve it with water, and transfer it to a 250mL volumetric flask. Dilute to the mark and shake well. Accurately pipette 25.00mL of KHP solution into the titration cup, and start the automatic potential titrator. Take the volume corresponding to the midpoint of the titration curve as the end point. Record the volume, remove the titration vessel, and rinse the burette and electrode with deionized water. Repeat the titration three times, and calculate the concentration of NaOH solution.

(3) **Determination of acetic acid concentration.** Accurately transfer 25.00mL of acetic acid solution into the titration cup with a pipette, and start the automatic potential titrator. Take the volume corresponding to the midpoint of the titration curve jump as the titration end point. Record the volume, remove the titration cup, and rinse the burette and electrode with deionized water. Repeat titration three times and calculate the concentration of acetic acid.

(4) **Powering off.** Cut off the power supply, clean the electrode and burette, dry the electrode with filter paper and put it back in the electrode box.

Data processing

Record the data of standardization of NaOH in Table 3-8. Record the data of HAc determination and its dissociation constant determination in Tables 3-9 and 3-10, respectively.

Table 3-8 Standardization of NaOH

item	1	2	3	Average
m_{KHP}/g				
V_{NaOH}/mL				
c_{NaOH}/(mol/L)				

Table 3-9 Determination HAc concentration

item	1	2	3	Average
V_{HAc}/mL	25.00	25.00	25.00	25.00
V_{NaOH}/mL				
c_{HAc}/(mol/L)				

Table 3-10 Determination HAc dissociation constant

item	1	2	3	Average
$1/2V_{ep,NaOH}$/mL				
pH at $1/2V_{ep,NaOH}$				
K_a				

Notes

Calomel electrode and glass electrode should be carefully used to prevent breakage. The glass electrode should be soaked in deionized water after use.

Questions

(1) Briefly describe the differences between chemical titration and potentiometric titration.

(2) List the steps for using a potentiometric titrator.

Exp. 26　Study of Ferrocyanide/Ferricyanide Redox Couple by Cyclic Voltammetry

Objectives

(1) To use an electrochemical workstation.

(2) To learn polishing the surface of a solid electrode.

(3) To learn to judge the reversibility of the electrode process.

Principles

Cyclic voltammetry (CV) is a versatile electroanalytical technique for the study of electroactive species. CV measures the current change with a triangular pulse voltage applied to the working electrode that varies linearly with time. First, the potential scans positively to the high value, and then it is scanned in reverse, causing a negative scan back to the original potential to complete the cycle. The resulting response of current changing with the applied voltage is the cyclic voltammogram.

A redox couple in which half reactions rapidly exchange electrons at the working electrode is said to be an electrochemically reversible couple. $[Fe(CN)_6]^{4-}/[Fe(CN)_6]^{3-}$ is a typical reversible redox system. As the potential is scanned positively (forward scan) and is sufficiently positive to oxidize $[Fe(CN)_6]^{4-}$, the anodic current is due to the following electrode process:

$$[Fe(CN)_6]^{4-} \longrightarrow [Fe(CN)_6]^{3-} + e^-$$

The electrode acts as an oxidant and the oxidation current increases. When the concentration of $[Fe(CN)_6]^{4-}$ at the electrode surface depletes, the current reaches a peak value and then decays gradually. As the scan direction is switched to negative, for the reverse scan the potential is still sufficiently positive to oxidize $[Fe(CN)_6]^{4-}$, so the anodic current continues even though the potential is now scanning in the negative direction. When the electrode becomes a sufficiently strong reductant (negative enough), $[Fe(CN)_6]^{3-}$, which has been forming adjacent to the electrode surface, will be reduced by the electrode process as below, resulting in a cathodic current.

$$[Fe(CN)_6]^{3-} + e^- \longrightarrow [Fe(CN)_6]^{4-}$$

Similarly, a peak cathodic current is obtained as $[Fe(CN)_6]^{3-}$ in the solution adjacent to the electrode is consumed, and then decays. Cyclic voltammogram records the anodic peak current i_{pa}, cathodic peak current i_{pc}, anodic peak potential E_{pa}, and cathodic peak potential E_{pc}, respectively.

For such a reversible couple, i_{pa} is equal to i_{pc} ($i_{pa}/i_{pc} = 1$). The difference between the peak potentials is described by the relationship $\Delta E = E_{pa} - E_{pc} \approx 0.059/n$, in which n is the number of electron transfer involved in the redox reaction. For slow electron transfers at the electrode surface, i. e. irreversible processes, the difference in peak potential increases and i_{pa} is not equal to i_{pc}, which can determine the reversibility of the electrode process.

For the diffusion-controlled reversible systems, the peak current is given by Randles-Sevcik equation,

$$i_p = 2.69 \times 10^8 n^{3/2} A D^{1/2} v^{1/2} c$$

where i_p is the peak current, n is the electron transfer number, A is the electrode area, D is the diffusion coefficient (m^2/s), c is the concentration of the measured substance (mol/L) and v is the scan rate (V/s).

In this experiment, we prepare different concentrations of the $K_3Fe(CN)_6/K_4Fe(CN)_6$ standard solution and investigate the relationship between peak current and concentration/scan rate in cyclic voltammetry. Meanwhile, evaluate the electrochemical behavior of a reversible couple.

Equipment and reagents

(1) **Equipment.** CHI 660E electrochemical workstation (Shanghai Chenhua), gold electrode, platinum wire electrode, saturated calomel electrode, and electrode polishing material.

(2) **Reagents.** 0.30mol/L $K_3Fe(CN)_6/K_4Fe(CN)_6$ stock solution (containing 1.0mol/L potassium chloride).

Procedures

(1) **Preparation of standard solutions.** In each of the five 50mL volumetric flasks, add 3.00mL, 5.00mL, 6.00mL, 7.00mL, and 8.00mL of $K_3[Fe(CN)_6]/K_4Fe(CN)_6$ stock solution (0.30mol/L, containing 1.0mol/L potassium chloride), respectively. Then dilute to the mark with 1.0mol/L potassium chloride solution, and mix well.

(2) **Polishing of the working electrode.** Polish the working electrode to a mirror surface with Al_2O_3 powder (particle size 0.05μm), clean it with deionized water, and dry it.

(3) **Setting operating parameters**

① Turn on the CHI 660E electrochemical workstation and computer for preheating (10min).

② Fill the cell with a standard solution and insert the electrodes. Use the newly treated gold electrode as the working electrode, the platinum wire electrode as the counter electrode, and the saturated calomel electrode as the reference electrode (connecting the red, green and white clip to Pt auxiliary electrode, gold electrode and saturated calomel reference electrode, respectively).

③ Click Setup manual, select the Cyclic Voltammetry method in the Technique item and set the parameters suggested below (which may be changed appropriately).

Initial potential (Init E): $-0.2V$

Highest potential (High E): $+0.6V$

Lowest potential (Low E): $-0.2V$

Final potential (Final E): $+0.6V$

Scan rate: 0.1 V/s

Sample interval: 0.001V/s

The polarization time at the initial potential or the stopping time (Quiet time): 5s

Sensitivity: 0.001

After completing the above parameter settings, click the OK button after double check. Then click the Run button. Now, the instrument starts the cyclic voltammetry scan and records the cyclic voltammogram. After the measurement, save the cyclic voltammogram.

④ Cyclic voltammograms with different concentrations. Record cyclic voltammograms of each concentration of $K_3Fe(CN)_6$ / $K_4Fe(CN)_6$ solutions prepared in the potential range from $-0.2V$ to $0.6V$ at a scanning rate of 100mV/s.

⑤ Cyclic voltammograms at different scan rates. Use one standard solution and scan the cyclic voltammograms from -0.2 to $0.6V$ at the following scan rates (v): 50mV/s, 80mV/s, 100mV/s, 150mV/s and 200mV/s.

Data processing

(1) Record the i_{pa}, i_{pc}, E_{pa}, and E_{pc} values from the cyclic voltammogram of the $K_3Fe(CN)_6$ / $K_4Fe(CN)_6$ solution.

(2) Discuss the effect of scan rate on i_p by plotting i_{pc} vs $v^{1/2}$ and i_{pa} vs $v^{1/2}$.

(3) Describe the effect of scan rate on ΔE_p.

(4) Plot i_{pa} and i_{pc} vs. the concentration of $K_3Fe(CN)_6$, respectively, describe the relationship between the concentration and peak current.

(5) Discuss the reversibility of $K_3Fe(CN)_6$ / $K_4Fe(CN)_6$ redox couple in KCl solution according to your results.

Notes

(1) Before use, the gold electrode should be carefully polished, ultrasonically washed and rinsed with deionized water. The treatment should be patient and meticulous, otherwise, the experimental results will be seriously affected.

(2) To make the liquid phase mass transfer process be controlled only by diffusion, electrolysis should be carried out in the static solution with added electrolyte.

(3) In order to restore the initial surface of the electrode surface for different scans, the electrode should be lifted and then immersed in the solution, or the solution should be stirred and kept stationary before use.

(4) Avoid short circuit caused by electrode contact.

Questions

(1) Why should the electrode surface be cleaned before the experiment?

(2) Why should we keep the solution still during the scanning?

Exp. 27   Determination of Vitamin C in Fruit Juice Beverages by Differential Pulse Voltammetry

Objectives

(1) To master the basic principles of differential pulse voltammetry.

(2) To master the method of measuring vitamin C content by differential pulse voltammetry.

Principles

Differential pulse voltammetry (DPV) is a highly sensitive voltammetric analysis technique. It superimposes a constant amplitude pulse voltage on a slowly varying direct current (DC) voltage. The pulse amplitude (height) is generally ten to several tens of millivolts, and the duration is 40 to 60 milliseconds. It records the current difference between the end of pulse and that before the pulse. Due to the two-time current sampling, the background current caused by the DC voltage can be well subtracted. Thus, the sensitivity of DPV is higher than the cyclic voltammetry (CV). The DPV peak current is proportional to the magnitude of the pulse amplitude and is not affected by the residual current. The peak current of DPV is proportional to the concentration of the substance and can be used for quantitative analysis.

Vitamin C, also known as ascorbic acid (Vc), is a chemical substance essential to the human body. It loses electrons and is oxidized at the electrode surface, generating an oxidation current. The content of Vc in real samples such as vitamin C juice drinks or vitamin tablets can be determined by the standard curve method.

Equipment and reagents

(1) **Equipment.** CHI 660E electrochemical workstation (Shanghai Chenhua), pH meter, ultrasonic cleaner, saturated calomel electrode, platinum electrode, glassy carbon electrode, electrode polishing material, 100mL pipette, 10mL pipette, beaker, glass rod, and dropper.

(2) **Reagents.** Vitamin C (AR, $C_6H_8O_6$, $M_r = 176g/mol$), vitamin C juice drink,

deionized water, ethanol, KCl, KNO$_3$, HCl, NaOH, phosphate buffer with pH = 4.00, 50mmol/L K$_3$Fe(CN)$_6$ solution (containing 0.5mol/L KCl), 500mL 1 : 1 HNO$_3$ solution.

Procedures

(1) Pretreatment of glassy carbon electrode. Polish the electrode with a small amount of α-Al$_2$O$_3$ powder on the suede, and then rinse the surface with deionized water. If necessary, ultrasonically clean it with 1 : 1 ethanol, 1 : 1 HNO$_3$, and deionized water successively, 2~3 minutes each time.

Pipette 15mL of K$_3$Fe(CN)$_6$ solution into the electrolytic cell and insert a clean three-electrode system. Scan the CV from 0.6 V to −0.2V at a scanning rate of 50mV/s. If the peak potential difference is within 100mV, the electrode can be subjected to the following related experiments.

(2) Plotting the calibration curve

① Weigh an appropriate amount of vitamin C (~0.22g), and dissolve it in a phosphate solution of pH = 4.00. Transfer the solution to a 50mL brown volumetric flask, dilute to the mark and mix well (about 25mmol/L).

② Accurately transfer 0.50, 1.00, 1.50, 2.00, and 2.50mL of the as-prepared Vc standard solution to the electrolytic cell, respectively, and add an appropriate amount of phosphate buffer solution (the total volume is about 10mL). Under the optimal experimental conditions, record the *i-E* curve by DPV technique with GCE as the working electrode, SCE as the reference electrode, and platinum electrode as the auxiliary electrode. After the end of each scan, the electrode should be removed and rinsed with deionized water, and then dried with the filter paper before the next measurement, to maintain the stability and reproducibility of the modified electrode. Construct the calibration curve of Vc by plotting the peak current against the concentration of Vc standard solutions.

(3) Sample determination

Pipette 10.00mL of the juice sample to a 50mL volumetric flask, dilute to the mark with phosphate buffer solution and mix well. Take 10mL of the above solution for the test, and repeat three times to obtain the DPV curves.

Data processing

(1) Construct the calibration curve of Vc by plotting the peak current against the concentration of Vc standard solutions and obtain the linear regression equation.

(2) Calculate the content of vitamin C in the Vc juice drinks.

Questions

(1) Why can differential pulse voltammetry achieve higher sensitivity?

(2) Why do the differential pulse voltammograms show peaks?

Exp. 28 Determination of Trace Lead and Cadmium in Water Samples by Anodic Stripping Voltammetry

Objectives

(1) To master the basic principles of anodic stripping voltammetry.

(2) To perform the anodic stripping voltammetry in the electrochemical workstation.

(3) To master the standard addition method for quantitative analysis.

Principles

Anodic stripping voltammetry (ASV) involves two basic steps. First, the analytes are concentrated onto the working electrode at an applied negative potential. Then the enriched analytes are stripped from the electrode. The content of the analyte is determined according to the dissolved peak current.

ASV is one of the most sensitive, convenient and cost-effective analytical methods for the determination of metal ion contaminants in water (e. g. rivers, lakes, drinking water sources, etc.). With ASV, it is possible to analyze very low levels of several metals such as Pb, Cd, Cu and Zn, simultaneously. The detection sensitivity of ASV can reach the level of part-per-million (mg/L) or even part-per-billion (μg/L).

As for trace Pb and Cd determination in water samples, the electrode potential can be controlled to -1.0V (vs. SCE) in an acidic condition, so that Pb^{2+} and Cd^{2+} could be simultaneously enriched on the working electrode (self-made glassy carbon mercury membrane electrode) in their metallic form. Then the anode is scanned linearly to -0.1V (vs. SCE) to obtain two reoxidizing peaks, in which the stripping peak around -0.4V (vs. SCE) is ascribed to Pb, while -0.6V (vs. SCE) to Cd. As the current of the stripping peak is proportional to the concentration of Pb^{2+} and Cd^{2+} in the solution, it can be used for quantitative analysis of Pb and Cd, respectively.

Generally, the standard curve method or standard addition method can be used for quantitative determination. The calculation formula of the standard addition method is as follows:

$$c_x = \frac{c_s V_s i_x}{i(V_x + V_s) - i_x V_s}$$

Where c_x, V_x, and i_x are the concentration, volume, and peak current of the stripping peak, respectively. c_s and V_s are the concentration and volume of the added standard solution, respectively, while i is the stripping peak current measured after the addition of the standard solution.

Equipment and reagents

(1) **Equipment.** CHI 660E electrochemical workstation, glassy carbon electrode (GCE),

platinum wire electrode, saturated calomel electrode, electrolytic cell (25mL), magnetic stirrer with a Teflon-coated stirring bar, and nitrogen gas cylinder.

(2) Reagents. 1.000mg/mL Pb^{2+} standard solution, 1.000mg/mL Cd^{2+} standard solution, 0.02mol/L $HgSO_4$ solution, 2.0mol/L HAc-NaAc buffer solution (pH = 5.0).

Procedures

(1) Preparation of glassy carbon mercury film electrode. Polish the GCE as the methods described in Exp. 21. Add 10mL of deionized water and $100\mu L$ of $HgSO_4$ solution to the electrolytic cell. Insert the three-electrode system and then connect them to the electrochemical workstation. Set the electrode potential to $-1.0V$ (vs. SCE), and then electroplate for 5 minutes under a nitrogen atmosphere to prepare the glassy carbon mercury film electrode.

(2) Determination of Pb and Cd peak potential

① Set the parameters of the electrochemical workstation as below: Linear scan mode, initial potential at $-1.2V$, final potential at $+0.5V$, scan rate to 100mV/s.

② Add 10.00mL of deionized water and 1.00mL of HAc-NaAc solution into the electrolytic cell, deoxygen with nitrogen for 10min, insert the three-electrode system, turn on the stirrer, and electrolytically enrich for 60s (enrichment potential $-1.2V$). Then turn off the stirrer and stop the enrichment. After standing for 30s, start scanning from $-1.2V$ to $+0.5V$ and record the blank stripping curve.

③ Add 20.0μL of 1.000mg/mL Pb^{2+} standard solution and 200μL of 1.000mg/mL Cd^{2+} standard solution to the above blank solution, repeat step②, and record the stripping curve. After the measurement, clean the three-electrode system under$+0.1$ V for 30s.

④ Increase the adding amount of Pb^{2+} and Cd^{2+}, and change the experimental conditions such as enrichment time, scan rate and enrichment potential to observe the change of the stripping curve. Find out the peak potential of Pb^{2+} and Cd^{2+}, which can be used as the criteria for qualitative analysis.

(3) Quantitative determination

① Add 10.00mL of water sample and 1.00mL of HAc-NaAc solution to the electrolytic cell. Record the stripping voltammetry curve as in step (2), and repeat it twice.

② Add a certain amount of Pb^{2+} and Cd^{2+} standard solution to the above electrolytic cell (depending on the amount of ions to be measured in the water sample). Record the stripping voltammetry curve as in step (2), and repeat it twice.

Data processing

Calculate the content of Pb and Cd in the water sample according to the calculation formula of the standard addition method from the two peak current values of the stripping peak before and after adding the standard solution.

Notes

(1) If the reagent blank value is large, the blank value should be deducted to avoid

large errors when calculating the content.

(2) As the test solution contains mercury, it can only be collected into the designated recycling bottle. It is strictly forbidden to pour into the sink to avoid environmental pollution.

Questions

(1) Why is anodic stripping voltammetry more sensitive?

(2) What are the factors that affect anodic stripping voltammetry? How should it be controlled?

Exp. 29   Determination of Ascorbic Acid in Vitamin C Tablets by Coulometric Titration

Objectives

(1) To be familiar with the use of coulombs and related operating techniques.

(2) To learn and master the basic principles of coulometric titration for the determination of ascorbic acid.

Principles

Vitamin C (ascorbic acid) is one of the most important vitamins for maintaining human health. The methods for determining the content of vitamin C include iodometry, ultraviolet spectrophotometry, voltammetry, infrared spectroscopy and coulometric titration. Among them, coulometric titration does not need to prepare and calibrate standard solutions, and the analysis results are obtained by accurately determining the amount of electricity or potential. Therefore, it has the characteristics of high sensitivity, good precision and high accuracy.

Coulometric titration is an analytical method in which the titrant produced by electrolysis (galvanostatic or potentiostatic) reacts quantitatively with the substance to be measured in the electrolytic cell to determine the amount of substance. If the efficiency of electrolysis is 100%, the reaction of the electrically generated titrant with the measured substance is complete, and also there is a sensitive method to determine the end point, then the amount of electricity consumed is proportional to the amount of the measured substance. Quantitative calculations can be performed according to Faraday's law, which is as follows:

$$m = QM/nF$$

m —the mass of the substance being titrated;

Q —the amount of electric charge consumed by the electrode reaction;

M — the molecular weight of the analyte;

F — Faraday constant (96487C/mol);

n —the moles of electrons transferred.

In this experiment, a KLT-1 universal coulomb meter with 4 Pt electrodes (two for electrolysis reaction and two for end point signaling) was used to electrolyze KBr acidic solution with constant current and the resulting electrolytic reaction is as follows:

$$\text{Anode:} \quad 2Br^- = 2e^- + Br_2$$
$$\text{Cathode:} \quad 2H^+ + 2e^- = H_2 \uparrow$$

Br_2 generated at the Pt anode reacts with ascorbic acid according to the following reaction:

When ascorbic acid is consumed, the concentration of Br_2 suddenly rises, signaling the end of the reaction. The rise in Br_2 concentration is detected by measuring the current between the two detector Pt electrodes. A voltage of about 150mV applied between these two electrodes is not enough to electrolyze any solute, so only a tiny current of less than $1\mu A$ ows through the microammeter. At the equivalence point, ascorbic acid is consumed, $[Br_2]$ suddenly increases, and the pointer of the microampere meter is deflected by virtue of the reactions:

$$\text{Detector anode:} \quad 2Br^- = 2e^- + Br_2$$
$$\text{Detector cathode:} \quad Br_2 + 2e^- = 2Br^-$$

The content of the analyte can be calculated by the amount of charge consumed and the number of electrons transferred.

Equipment and reagents

(1) **Equipment.** KLT-1 universal coulomb meter, magnetic stirrer, electrolytic cell (double platinum working electrode, double platinum indicating electrode), pipette (1mL), cylinder (100mL).

(2) **Reagents**

① Electrolyte: 1 : 2 HAc is mixed with 0.5mol/L KBr solution in equal volume. ②Ascorbic acid standard solution (0.5g/L): accurately weigh 0.05g of the solid Vc (standard) and dissolve it in deionized water. Transfer the solution to a 100 mL volumetric flask. Add water to the mark and mix well. ③Sample solution: accurately weigh one vitamin C tablet in a small beaker, soak it in a small amount of deionized water for a while, carefully mash it with a glass rod, and dissolve it in an ultrasonicator. After the tablet is dissolved (a small amount of excipients in the tablet is insoluble), transfer the solution with the residue to a 50mL volumetric flask and dilute to the mark with deionized water.

Procedures

(1) Turn on the power switch according to the instruction manual of the instrument,

and preheat it for 20-30 minutes.

(2) Add 50mL of KBr electrolyte into the electrolytic cell and place a stir bar. Use a dropper to take the electrolyte into the working cathode cannula to raise it above the external liquid level. Insert the cleaned electrode into the solution, place the electrolytic cell on the magnetic stirrer and fix it with the clip. Turn on the stirrer, and adjust the appropriate speed.

(3) Pre-electrolysis: Select the range to 5mA, set the current fine-tuning (clockwise) to the maximum position first, and press the "current" (the indicator electrode should be clamped on the two platinum indicator electrodes), "up", "start", and "polarization potential" button, successively. Adjust the polarization potential knob to set the polarization potential about 150mV and then release the "Polarization potential" button. Press the "electrolysis" button, the red light is off. Switch the "work/stop" button to "work" for pre-electrolysis. When the electrolysis reaches the ending point, the needle deflects rapidly to the right with the red light on, indicating that the electrolysis is terminated.

(4) Release the "Start" button. Add 1.00mL of ascorbic acid standard solution into the electrolysis cell, insert the electrode and press the "start" button. Press the "electrolysis" button to perform electrolysis, and record the amount of electricity consumed (millicoulomb) at the end point. Repeat the measurement twice. If there is too much solution in the electrolytic cell, pour out part of the solution for use. (Note: use different pipettes for the standard solution and the sample solution)

(5) Measure the sample solution twice as in step (4).

(6) After measurement, clean the electrolytic cell and the electrodes, then inject deionized water.

Data processing

(1) Calculation of current efficiency:

$\eta = Q_{standard}/Q_{measurement} = [(m/M)nF]/Q_{measurement} = (CV)_{std}\, nF/(1000Q_{measurement})$.

(2) Record the data in Table 3-11 and calculate the Vc content in the tablets.

$$m = M\eta Q_{measurement}/(nFm_{sample})$$

Table 3-11　The results of Vc determination

Number of measurements (n)	Mass of Vc tablet(m_{Vc})/g	Charge consumed/mC	Content of Vc /(mg/g)		
			Each measure	Average	RSD/%

Questions

(1) What is the effect of adding KBr and glacial acetic acid to the electrolyte?

(2) If KBr used is oxidized by O_2 in the air, what effect will it have on the results?

(3) During electrolysis, how does the continuous appearance of H_2 in the cathode affect the pH value of the electrolyte?

(4) Why does the cathode need to be placed in the protective sleeve, while the end point indicator electrode does not?

Exp. 30 Conductometric Titration of Mixed Acids

Objectives

(1) To master the basic principles and experimental methods of measuring the concentration of mixed acid by conductometric titration.

(2) To be familiar with the components, performance and usage of DDS-11A conductivity meter.

Principles

At a certain temperature, the conductivity of the electrolyte solution is related to the ion composition and concentration in the solution, which change constantly during titration. Therefore, changes in conductivity can be used to indicate the end point of the reaction. Conductometric titration is a titration analysis method that continuously observes the electrolytic conductivity of the reaction mixture when a reactant is added and determines the end point via the change of conductivity before and after the titration end point. In the titration of the mixture of HCl and HAc versus NaOH solution, HCl gets neutralized first. H^+ with greater mobilities in the solution is neutralized by adding OH^- to produce water that is difficult to ionize and Na^+ with fewer mobilities. The reaction is as follows:

$$HCl + NaOH \stackrel{}{=\!=\!=} H_2O + NaCl$$

The molar conductance of Na^+ is less than that of H^+ ($\lambda_{\ominus,Na^+} = 50.0 \times 10^{-4} \Omega^{-1}$, $\lambda_{\ominus,H^+} = 341.8 \times 10^{-4} \Omega^{-1}$). Therefore, before the stoichiometric point, the conductivity of the solution gradually decreases with the addition of NaOH. After the stoichiometric point, the excessive hydroxide ions with greater mobilities would increase the conductivity of the solution.

When HCl is completely neutralized, the weak acid CH_3COOH begins to be neutralized to generate water that is difficult to ionize and NaAc which is easy to dissociate. The reaction is as follows:

$$CH_3COOH + NaOH \stackrel{}{=\!=\!=} CH_3COONa + H_2O$$

Where the conductivity is first raised as weak acid is converted to its salt, and then raises rapidly as the excess alkali is introduced.

Taking the conductance of the solution as the ordinate and the volume of the titrant sodium hydroxide as the abscissa, a titration curve with two inflection points can be obtained. According to the volume of sodium hydroxide at the inflection point, the respective contents of HCl and HAc can be calculated.

Equipment and reagents

(1) **Equipment.** DDS-11A conductivity meter, magnetic stirrer, DJS-2 platinum black conductivity electrode, basic burette, beakers, pipette, and pipette ball.

(2) **Reagents.** 0.1mol/L NaOH standard solution, 1% phenolphthalein indicator, mixed acid with an unknown concentration, potassium hydrogen phthalate (KHP, AR).

Procedures

(1) **Standardization of sodium hydroxide standard solution.** Using KHP as the primary standard, measure the exact concentration of NaOH standard solution.

(2) **Titration of mixed acids**

① Accurately transfer 25.00mL of the mixture into a 250mL beaker and then dilute to about 100mL using deionized water.

② Drop a stirring bar into the beaker and insert the conductivity electrode into the solution.

③ Start the magnetic stirrer and adjust the rate at a suitable speed. Record the conductivity value.

④ Start addition of the NaOH titrant and record the conductivity value after each addition of 0.5mL increments till about 30mL of the base is added.

⑤ Repeat steps from ① to ④ and record the results.

⑥ Plot the graph concerning the volume of NaOH consumed versus conductivity.

⑦ Calculate the amount of HCl and HAC present in the mixed acid.

Data processing

(1) According to the standardization results of sodium hydroxide solution, calculate the concentration of NaOH solution according to the following equation.

$$c_{\text{NaOH}} = \frac{m_{\text{KHP}}}{M_{\text{KHP}} V_{\text{NaOH}}} \cdot 1000 \ (\text{mol/L})$$

(2) Plot the titration curve using the conductivity values as the y-axis and the volume of titrant added as the x-axis. According to the point of intersection of the 3 curves, calculate the amount of HCl and HAC present in the mixed acid.

Questions

(1) Explain the difference between the conductivity titration curves of hydrochloric acid and acetic acid titrated with sodium hydroxide. Why?

(2) What are the advantages and drawbacks of conductometric titration?

Exp. 31　Determination of the Composition and Content of n-Alkanes by Gas Chromatography

Objectives

(1) To learn the basic structure and operating method of gas chromatograph.

(2) To master the peak area normalization method in quantitative analysis.

Principles

Gas chromatography (GC) is a chromatographic separation analysis technique in which gases are mobile phases. In gas chromatography, the vaporized analyte is transported through the heated column by the carrier gas to achieve separation. Qualitative analysis is performed based on the position (retention time) of the peak on the chromatographic elution curve (chromatogram), and quantitative analysis is done according to the area or height of the peak. There are three widely used quantitative methods: external standard method, internal standard method and normalization method. The purpose of this experiment is to determine the composition and content of the mixed alkanes. In this experiment, the retention time of a standard sample is used for qualitative analysis, and the peak area normalization method without a correction factor is used for quantitative analysis.

The basis of quantitative analysis is that under certain conditions, the mass m of the measured substance is proportional to the response value of the detector, namely:

$$m_i = f_i A_i \text{ or } m_i = f_i h_i$$

where A_i represents the peak area of the measured components, h_i stands for the peak height, and f_i is the correction factor.

Since the amount of a component is proportional to its peak area, if all components in the sample can generate signals and obtain corresponding chromatographic peaks, the following normalization formula can be used to calculate the content of each component. That is to calculate the sum of the contents of all components in the sample by 100%, then find the peak area and correction factor of each component in the sample respectively, and then calculate the percentage content of each component in turn.

$$\omega_i = \frac{A_i f_i}{\Sigma A_i f_i} \times 100\%$$

If the correction factor of each component in the sample is similar, the correction factor can be eliminated, and the peak area normalization can be directly used for calculation.

Equipment and reagents

(1) **Equipment.** 7820A gas chromatograph with hydrogen flame ionization detector (FID). SE-54 column (30m × 0.32mm × 0.33μm), 1μL microsyringe.

(2) Chromatographic conditions

① Carrier gas: N_2, 34mL/min. ② Inlet temperature: 220℃; split ratio: 1 : 30. ③FID detector parameters: temperature is 250℃, the hydrogen flow rate is 30mL/min, and air flow rate is 300mL/min. ④Temperature-programming: first raise the column temperature to 80℃ and maintain 1min, and then heat up to 180℃ at 25℃/min and maintain 1min.

(3) Reagents. Alkane mixtures (acetone as solvent), alkane standards (n-heptane, n-octane, n-nonane, n-decane).

Procedures

(1) Startup and parameter settings

① Open the carrier gas (N_2), and turn on the GC machine and computer. Turn on the heating switch of the column oven, detector, and sampler inlet on the main interface.

② Establish the analytical method and set the chromatographic conditions.

③ After the baseline is stable, the machine is ready for sample injection.

(2) Sample analysis and determination

① Accurately pipette 0. 4μL of each alkane standard with a 1μL microsyringe, inject them into the instrument sequentially, and record their t_R (min) and peak area (A).

② Under the same chromatographic conditions, pipette 0. 4μL of alkane mixture with a 1μL microsyringe, inject it into the instrument, and record its t_R (min) and peak area (A).

(3) Powering off

① Turn off the column oven, detector, and injector heating switch on the main interface of the chromatography workstation.

② Turn off hydrogen and air.

③ After the detector temperature drops to 50℃, turn off the GC machine.

④ Turn off the carrier gas.

Data processing

(1) According to the t_R value of each alkane standard and the specific value in the mixture, assign the peaks in the alkane mixture sample.

(2) According to the peak area of each component, calculate the percentage content of each component by the normalization method.

Questions

(1) What are the quantitative methods of chromatography? Describe their scope of application.

(2) What are the requirements for injection?

Exp. 32　Quantification of Thymol in Mouthwash by Gas Chromatography

Objectives

　　(1) To master the quantification method of internal standard.

　　(2) To master the method of determining thymol concentration with GC.

Principles

　　Thymol is a naturally occurring biocidal agent. It has very strong antiseptic properties and is one of the ingredients in mouthwashes. Gas chromatograph (GC) with a hydrogen flame ionization detector (FID) is used for separation and determination of thymol in the mouthwash, and quantitative analysis is performed by the internal standard method.

　　An internal standard method is to add a known amount of the internal standard, different from the analyte, to the unknown. Signal from analyte is compared with signal from the internal standard to find out how much analyte is present. This method is especially useful for analyses in which the quantity of sample analyzed or the instrument response varies slightly from run to run.

　　In this experiment, in the series of standard solutions of thymol, a certain amount of the internal standard tetradecane is added sequentially, and the series of standard solutions containing the internal standard are injected into the column for chromatographic analysis. Taking the ratio of peak area of thymol to tetradecane as the y-axis, the concentration of thymol standard solution as the x-axis, plot the standard curve, and get a linear regression equation. Then, under the same chromatographic conditions, the mouthwash containing the internal standard is also analyzed, and the ratio of the peak area of thymol to tetradecane in mouthwash is substituted into the regression equation, and the content of thymol in mouthwash can be calculated.

Equipment and reagents

　　(1) **Equipment.** 7820A gas chromatograph with hydrogen flame ionization detector (FID). HP-INNOWax (polyethylene glycol) column (30m × 0.32mm × 0.25μm), 1μL microsyringe, 10mL volumetric flask.

　　(2) **Chromatographic conditions.** Carrier gas: N_2, 1.5mL/min. Inlet temperature: 240℃, splitless injection. FID detector parameters: temperature is 260℃, the hydrogen flow rate is 40mL/min, airflow rate is 400mL/min. Temperature-programming: first raise the column temperature to 100℃ and maintain 2min, and then heat up to 240℃ at 10℃/min and maintain 5min.

　　(3) **Reagents.** Methanol, 1000mg/L thymol solution, mouthwash, tetradecane.

Procedures

(1) Construction of the internal standard curve

① From the 1000mg/L stock solution, prepare at least 5 standards in the 10mL volumetric flasks provided in the range of 100 to 500mg/L.

② Add 5μL of tetradecane to each flask and mix well. The tetradecane acts as an internal standard. Use methanol for diluting solutions.

③ Inject all the standards onto the GC, and record the retention times and areas.

(2) Sample determination

① Add 5mL of mouthwash to a 10mL volumetric flask. Add 5μL tetradecane to the solution. Dilute to volume with methanol. Mix well.

② Inject your diluted mouthwash sample onto the GC. Identify the thymol and tetradecane peaks. Record their peak areas.

Data processing

(1) Calculate the ratio of the thymol peak area to the tetradecane. Generate a calibration curve and the corresponding linear regression equation by plotting concentration vs. ratio of peak area.

(2) Calculate the ratio of thymol peak area to tetradecane peak area in mouthwash. Use the linear regression to work out the thymol concentration in the mouthwash and report the results in mg/L.

Questions

(1) Discuss sources of error in the experiment and estimate the uncertainty in the results obtained.

(2) Why do you think there are more peaks present in the sample?

Exp. 33   Determination of *m*-Nitrophenol Content in Water Samples by High-performance Liquid Chromatography

Objectives

(1) To understand the basic structure and operation skills of HPLC.

(2) To master the working principle of HPLC.

(3) To master the basic methods of qualitative and quantitative analysis of HPLC.

Principles

High-performance liquid chromatograph (HPLC) uses high pressure to force solvent through closed columns containing fine particles that give high-resolution separations. HPLC mainly consists of an autosampler, a solvent delivery system, a sample injection valve, a high-pressure chromatography column, and a detection and data processing system.

Reversed-phase chromatography is a commonly used HPLC technique in which the stationary phase is nonpolar or weakly polar and the solvent is more polar. The widely used stationary phase in reversed-phase HPLC is octadecyl bonded silica (ODS), and the mobile phase is water, methanol, acetonitrile, etc. When the partition coefficient (K) of the components in the mixture is different in the two phases, the components move at different speeds with the mobile phase to the exit of the column, resulting in separation. The component with a small K elutes out first, and the component with large K peaks later.

M-Nitrophenol is an extremely harmful environmental pollutant and is one of the pollutants in environmental water. In this experiment, the nitrophenol was separated by reversed-phase HPLC. The working curve was plotted according to the peak area of the series standard solution versus the concentration of m-nitrophenol, and the concentration of m-nitrophenol in the water sample is obtained according to the peak area and the working curve.

Equipment and reagents

(1) **Equipment.** Waters 600 high-performance liquid chromatograph, Waters 2996 PDA detector, Empower operating system, $100\mu L$ microsyringe, $20\mu L$ loop, $045\mu m$ membrane filter, 10mL graduated tube, 1mL pipette.

(2) **Chromatographic conditions.** Symmetry C_{18} column (150mm $\times$ 3.9mm $\times$ 5μm), the mobile phase is 7 : 3 methanol and water (volume ratio), the flow rate is 0.8mL/min, room temperature, and the detection wavelength is 271 nm.

(3) **Reagents.** M-nitrophenol stock solution 0.1mg/mL, water sample.

Procedures

(1) Accurately pipette 0.2mL, 0.4mL, 0.6mL, 0.8mL and 1.0mL of m-nitrophenol stock solution in a 10mL graduated tube, dilute to the mark with water, and mix well. The concentrations of this standard solution series are 2.0μg/mL, 4.0μg/mL, 6.0μg/mL, 8.0μg/mL, 10μg/mL, respectively.

(2) Turn on the vacuum degasser, pump, and detector of the machine. Turn on the computer. Edit the instrumental methods of pumps and detectors in your computer, name and save them.

(3) Establish the instrument method, enter the sample name, sample number and run time, etc.

(4) Flush the system with the mobile phase for 5 to 10 minutes and start the injection after the pressure and baseline have stabilized.

(5) Place the injection valve in the LOAD position and use a micro syringe to take approximately $60\mu L$ of a 2μg/mL standard solution (filtered with a 0.45μm membrane) into the injection valve. Now the solution is loaded into the injection loop.

(6) Shift the injection valve from the load (LOAD) position to the injection (INJECT) position. The injected solution elutes to the column alone with the mobile phase and

starts to separate. Record retention time and peak area.

(7) Repeat steps (5)-(6) as the standard solution concentration increases.

(8) Sample analysis is performed as in steps (5)-(6). Repeat the determination 3 times, recording their retention times and peak areas.

Data processing

Integrate the chromatographic peaks of the series of m-nitrophenol standard solutions. Plot the working curve with the peak area as the y-axis and the concentration as the x-axis. Obtain the linear regression equation and correlation coefficient. Substitute the peak area of the m-nitrophenol in the water sample into the regression equation, and calculate the concentration of m-nitrophenol in the water sample.

Questions

(1) What are the similarities and differences between normal-phase chromatography and reversed-phase chromatography?

(2) Explain the principle of m-nitrophenol determination by reversed-phase chromatography.

(3) What are the commonly used detectors for high-performance liquid chromatography?

(4) Why does the mobile phase have to be filtered with a membrane?

Exp. 34   Determination of Rutin in Locust Rice by HPLC

Objectives

(1) To Master the basic methods of qualitative and quantitative analysis of HPLC.

(2) To be familiar with the principle and operation skills of HPLC instrument.

Principles

The locust rice is the dried buds of the *Sophora japonica* L. family. It has clearing heat and anti-inflammatory effects. Locust rice mainly contains flavonoids, saponins, sterols and other components, of which rutin content is the highest. Rutin has the effect of regulating the permeability of capillary walls, is clinically used as a capillary hemostatic agent, and can be used as an adjunct therapy for hypertension. Rutin is soluble in solvents such as methanol and water and has a large absorption at 254nm. The molecular structure of rutin is shown in Figure 3-7.

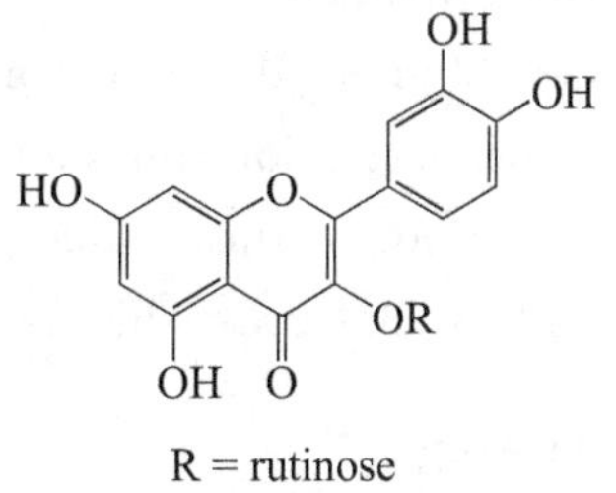

R = rutinose

Figure 3-7   The molecular structure of rutin

In this experiment, rutin and other components in locust rice are separated by reversed-phase chromatography and detected by ultraviolet spectroscopy. Under certain experimental conditions, the retention time and peak area are used for qualitative and quantitative analysis, respectively. Plot the peak area versus the concentration of the series standard solution of rutin, and obtain the calibration curve and linear equation. Find the content of rutin in the locust rice sample based on the peak area.

Equipment and reagents

(1) **Equipment.** Waters 600 HPLC, ultrasonicator, volumetric flask, pipette, stopper Erlenmeyer flask, microsyringe (50μL, flat head), membrane (0.45μm, organic).

(2) **Chromatographic conditions.** The column is C_{18} reverse-phase bonded phase column ($250\text{mm}\times4.6\text{mm}\times5\mu$m). The mobile phase is methanol and 0.5% aqueous glacial acetic acid solution (1:1, volume ratio), and the flow rate is 1.0 mL/min. The column temperature is 30℃. The detection wavelength is 254nm. The injection volume is 20μL.

(3) **Reagents.** Methanol (GR), acetic acid (AR), rutin standard, locust rice herb.

Procedures

(1) **Preparation of standard solutions**

① Take about 10mg of rutin standard, weigh it precisely, and add methanol to make 1mg/mL of rutin standard stock solution.

② Accurately pipette the standard stock solution 0.50mL, 1.00mL, 1.50mL, 2.00mL, 2.50mL in five 10mL volumetric flasks, add the mobile phase to the mark and shake well.

(2) **Preparation of sample solution**

① Take about 0.15g of locust rice powder, accurately weigh its mass, and put it into an Erlenmeyer flask with a stopper.

② Accurately add 25.00mL of methanol into the flask, weigh its mass and extract for 20 minutes in the ultrasonicator.

③ Cool down to room temperature and reweigh the accurate mass. Make up the weight lost with methanol, shake well, and then filter the solution.

④ Accurately transfer 5.00mL of the filtrate into a 100mL volumetric flask and dilute to the mark with the mobile phase.

(3) **Determination**

① Inject 20μL of different concentrations of rutin standard solution onto the column, and record their retention times and peak areas in Table 3-12, respectively.

② Inject 20μL of the sample solution, and record the retention time and peak area of rutin in Table 3-13. Repeat the assay 3 times.

Data processing

(1) Plot the standard curve with peak area versus the concentration of standard solu-

tions. Obtain the regression equation, correlation coefficient and linear range.

(2) Substituted the peak area of rutin into the regression equation and calculate the amount of rutin in locust rice.

Table 3-12   Peak area of different concentrations of rutin

Concentration of rutin/(mg/mL)	t_R/min	A
Regression equation		
Correlation coefficient		

Table 3-13   Calculation of rutin content in locust rice

Sample	Retention time, t_R/min	Peak area	Content of rutin/%	Average content/%
1				
2				
3				

Notes

The chromatographic conditions listed in the operation steps are reference data. Due to the different performances of the instruments and other factors, satisfying results may not be obtained. If necessary, appropriately adjust the concentration and ratio of the mobile phase to achieve the best separation.

Questions

(1) In this experiment, what is the effect of the change of mobile phase ratio on the retention time and separation effect of the rutin peak?

(2) When determining the rutin content in locust rice, why does the sample concentration have to fall within the standard curve?

Exp. 35   Determination of Cephalexin Capsules by HPLC

Objectives

(1) To master the method for the determination of cephalexin capsules by HPLC.

(2) To learn how to calculate the drug content using the external standard method.

Principles

Cephalexin is $(6R, 7R)$-3-methyl-7 $[(R)$-2-amino-2-phenylacetamido]-8-oxo-5-thia-1-azabicyclo $[4.2.0]$ oct-2-ene-2-carboxylic acid-hydrate. Its molecular formula is $C_{16}H_{17}N_3O_4S$ with a molecular weight of 347.39. Its chemical structural formula is shown in Figure 3-8.

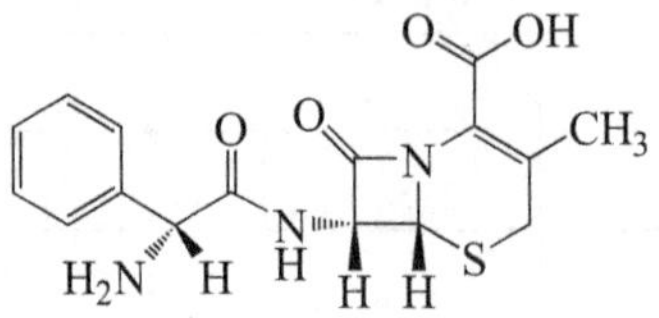

Figure 3-8   Molecular structure of cephalexin

In this experiment, the content of cephalexin capsules was determined by HPLC using external standard method. The sample containing cephalexin $(C_{16}H_{17}N_3O_4S)$ should be within 90.0%-110.0% of the marking amount.

Equipment and reagents

(1) Equipment. Shimadzu LC-20AD high-performance liquid chromatograph with UV detector, analytical balance, pipette, and volumetric flask.

(2) Chromatographic conditions. C_{18} reversed-phase bonded phase column (250mm × 4.6mm × 5μm) with water-methanol-3.86% sodium acetate solution-4% acetic acid solution (742 : 240 : 15 : 3) as a mobile phase, flow rate of $0.7 \sim 0.9$ mL/min, detection wavelength of 254 nm, injection volume of 20μL.

(3) Reagents. Cephalexin capsules, cephalexin reference substance, methanol (AR), sodium acetate (AR), acetic acid (AR), ultrapure water.

Procedures

(1) Preparation of test solution

① Accurately weigh 10 capsules of cefalexin. Remove the capsules and accurately weigh their mass. Calculate the average filling volume of one capsule.

② Mix the fillings evenly, weigh the appropriate amount (about equivalent to 0.1g of cephalexin) and place it in a 50mL beaker. Add the appropriate amount of mobile phase, and shake sufficiently to dissolve the cephalexin.

③ Transfer the dispersion to a 100mL volumetric flask, then dilute with the mobile phase to the mark, shake well, and filterate. Pipette 10mL of the filtrate into a 50mL volumetric flask, dilute to the mark with mobile phase, and shake well for further use.

(2) Preparation of reference solution. Accurately weigh a certain amount of the cephalexin reference substance to prepare a solution containing about 200μg of cephalexin per 1mL solution.

(3) Determination

① According to the operation steps of Shimadzu LC-20AD HPLC, start the machine, open the software, and set the separation parameters. Open the purge valve to exhaust bubbles, and wait for the baseline to be stable before injection.

② Accurately inject $20\mu L$ of the test solution and the reference solution into the liquid chromatograph respectively, record the chromatogram, and calculate the content by peak area according to the external standard method.

$$\text{Labeled percent amount} = \frac{A_x \times c_R \times D \times 10^{-3} \times \overline{W}}{A_R \times W \times \text{labeled amount of a grain(mg/grain)}} \times 100\%$$

Where A_x, A_R is the peak area of cephalexin in the test solution and the reference solution, respectively; D is the diluted volume of the test product; c_R is the concentration of the reference solution ($\mu g/mL$); W is the weight of the sample (g); $\overline{W}$ is the average grain weight (g/grain).

Notes

(1) The mobile phase and test solution need to be filtered with a membrane and degassed before use.

(2) Inject the reference and test solution at least twice and find their average value.

Questions

(1) Describe some other methods for the determination of cephalosporin antibiotics other than HPLC assay.

(2) Describe the principle and characteristics of the external standard method.

Exp. 36　Determination of Caffeine and Theophylline in Green Tea Beverages by HPLC

Objectives

(1) To learn the basic structure and operation of HPLC.

(2) To understand the principles, advantages and applications of reversed-phase liquid chromatography.

(3) To master the qualitative and quantitative methods in HPLC.

Principles

Green tea drink is a kind of beverage made from green tea powder. It has become one of the favorite drinks of the public with its excellent taste. Green tea drinks contain tea polyphenols, caffeine, theophylline, sucrose and other ingredients. Caffeine and theophylline are important bioactive substances that excite the cerebral cortex and eliminate fatigue. But drinking too much can cause some damage to the human body. Both caffeine and theophylline belong to the natural xanthine derivatives, and their chemical names are 1,3,7-trimethylxanthine and 1,3-dimethylxanthine, respectively.

The traditional methods for the quantitative determination of caffeine and theophylline are titration and UV/Vis spectrophotometry. In this experiment, the caffeine, theophylline, and other components were separated by reversed-phase chromatography and detected by the internal diode array detector. Under constant experimental conditions, the retention time (t_R) of the substance on the chromatogram is used as the qualitative parameter, and the peak area (A) is used as the quantitative parameter. The working curve is obtained by plotting peak areas of caffeine and theophylline standard solutions against the different concentrations. According to the peak area of caffeine and theophylline in the unknown sample, the content of caffeine and theophylline in the beverage was determined by the working curve method (namely the external standard method).

Equipment and reagents

(1) **Equipment.** Shimadzu LC-20AD HPLC, diode array detector, 100mL and 10mL volumetric flasks, 1.5mL injection bottles.

(2) **Chromatographic conditions.** ODS (C_{18}) column (150mm × 4.6mm × 5μm), mobile phase 70% water +30% methanol (volume ratio). The flow rate is 1.0μL/min, the detection wavelength is 272nm, and the injection volume is 10μL.

(3) **Reagents.** Methanol, caffeine, and theophylline standards, green tea drinks.

Procedures

(1) **Preparation of stock solutions of caffeine and theophylline.** Accurately weigh 10mg of caffeine and dissolve it with the prepared mobile phase. Transfer it into a 100mL volumetric flask, add the mobile phase to the mark and mix well. Prepare the theophylline standard stock solution according to the same method.

(2) **Preparation of standard solutions of caffeine and theophylline.** Accurately transfer 0.1mL of caffeine standard stock solution to a 10mL volumetric flask and dilute the solution to the mark with the mobile phase. Prepare the standard solution of theophylline in the same way.

(3) **Preparation of mixed standard solution series.** Pipette 0.10mL, 0.20mL, 0.30mL, 0.40mL, 0.50mL of caffeine standard stock solution and an equal volume of theophylline standard stock solution into a 10mL volumetric flask, respectively, and dilute the solutions to the marks with the mobile phase. The concentration of the resulting mixed standard solutions is 1μg/mL, 2μg/mL, 3μg/mL, 4μg/mL, 5μg/mL, respectively.

(4) **Construction the working curve**

① According to the operation steps of Shimadzu LC-20AD HPLC, start the machine, open the software, and set the separation parameters. Open the purge valve to exhaust bubbles, and wait for the baseline to be stable before injection.

② Inject the caffeine and theophylline standard solutions onto the column to determine the respective retention times. Then, inject the mixed standard solution from the lowest to the highest concentration. Record the peak areas of caffeine and theophylline at different concentrations.

(5) Sample determination

① Degas the green tea beverage by ultrasonication for 10min, filter it through a 0.45μm membrane, and dilute it 50 times with the mobile phase for further use.

② Inject the green tea sample, and record the retention time and peak area of caffeine and theophylline in the drink.

(6) Powering off. Once the entire running is over, rinse the column with gradient elution. Then, turn off the instrument.

Data processing

(1) Find the peaks of the corresponding caffeine and theophylline in the chromatogram based on the retention time of the standard sample.

(2) Obtain the calibration curves by plotting the peak areas (A) against the mass concentrations (ρ, μg/mL) of the standard samples.

(3) According to the calibration curves, find the mass concentrations (μg/mL) of caffeine and theophylline in the sample drink.

Notes

(1) Before injection, the beverage sample must be degassed and filtered. Direct injection will shorten the life of the column.

(2) The sample and standard solutions need to be refrigerated.

Questions

(1) Explain the principles of determining caffeine and theophylline by reversed-phase chromatography.

(2) Describe the qualitative and quantitative methods in HPLC.

Exp. 37　Evaluation of the Performance of HPLC Column and Determination of Resolution

Objectives

(1) To grasp the calculation method of the theoretical plate number, theoretical plate height and tailing factor.

(2) To grasp the calculation of resolution based on a chromatogram.

Principles

There are different methods and merits to evaluate the performance of the column. The main indicators include the resolution of the column, the number of theoretical plates, the height of the theoretical plates, the peak symmetry, the stability and reproducibility of the sample determination in different pH media, and the sample load of the column. Here the theoretical plate number, theoretical plate height, peak symmetry and resolution are mainly introduced.

The plate theory models the column as a distillation column, that is, the column is composed of a series of continuous, equal-volume plates. The height of each plate is called the height of the theoretical plate (H). It is approximately the length of column required for one equilibration of solute between mobile and stationary phases. Suppose that the column length is L, then n is the number of times the solute equilibrium is obtained, that is the theoretical number of plates (n). Obviously, the larger the number of plates n, the smaller the plate height H, and the higher the column efficiency. The number of plates is an important indicator to measure the performance of the column. In this experiment, the theoretical plate number of benzene and toluene is tested to determine the level of column efficiency.

The thermodynamic properties of the column and the uniformity of the column filling will affect the symmetry of the chromatographic peaks. The symmetry of the chromatographic peaks can be measured by the symmetry factor (f_s) or the tailing factor (T), which should be between 0.95 to 1.05.

The resolution (R) is an indicator of the total separation efficiency of the adjacent two components in the column from the chromatographic peak, and the resolution of the adjacent two components should be greater than 1.5 to achieve complete separation.

Equipment and reagents

(1) **Equipment.** HPLC (Shimadzu LC-20AD), C_{18} reversed-phase bonded column (250mm×4.6mm×5μm), UV detector, 50μL microsyringe.

(2) **Chromatographic conditions.** The mobile phase is 80% methanol+20% water (volume ratio), the flow rate is 1.0mL/min, the detection wavelength is 254nm, and the column temperature is 30℃.

(3) **Reagents.** Benzene (AR), toluene (AR), methanol (GR), and deionized water.

Procedures

(1) Prepare methanol solutions of benzene and toluene at 1μg/mL as test solutions.

(2) Inject 20μL of sample solution under the above chromatographic conditions and record the chromatogram.

Data processing

(1) Based on t_R (retention time), $W_{1/2}$ (half-width) value of the chromatographic peaks of benzene and toluene, calculate theoretical plate numbers as follows:

$$n = 5.54 \left(\frac{t_R}{W_{1/2}} \right)^2$$

(2) According to the chromatographic peak, calculate the tailing factor (T) of each component as follows:

$$T = \frac{W_{0.05h}}{2d_1}$$

Where $W_{0.05h}$ is the peak width in 0.05 peak height, d_1 represents the distance between peak maximum and peak front.

(3) Calculate the resolution of benzene and toluene based on their peaks as follows:

$$R = \frac{2(t_{R_1} - t_{R_2})}{W_1 + W_2}$$

t_{R_1} and t_{R_2} are the retention times of benzene and toluene, W_1 and W_2 are the peak widths of benzene and toluene.

Notes

(1) The solvents used in HPLC need to be purified. The mobile phase can only be used after degassing.

(2) At the end of the experiment, the reversed-phase column needs to be rinsed with methanol for 20~30min to protect the column.

Questions

(1) When using inverted bonded phase columns, what should be the pH range of the mobile phase?

(2) Is the column efficiency the same represented by benzene and toluene in the same column?

Exp. 38　Analysis of the Composition of Fatty Acid in Rapeseed Oil by Gas Chromatography-Mass Spectrometry

Objectives

(1) To master the basic components and working principles of GC-MS.

(2) To master the use of the NIST (National Institute of Standards and Technology) database.

(3) To identify the main fragment ion peaks in the mass spectrum and analyze the structure of typical organisms.

(4) To learn about the derivatization and extraction methods of multi-component mixed fatty acids, and the choice of chromatographic separation conditions.

Principles

Molecular mass spectrometry ionizes sample molecules by high-energy ion beams to generate various types of charged ions. These ions are separated by the electric field and magnetic field according to the mass-to-charge ratio (m/z) and are arranged into mass spectrometry. Qualitative and structural analysis of substances is performed according to the position of the peak of mass spectrometry, and quantitative analysis is done according to the intensity of the peak. Mass spectrometry has high sensitivity and strong qualitative specificity. Chromatography is characterized by efficient separation of mixtures and easy quantitative analysis. Combining these two analytical techniques can achieve separation and identification simultaneously, which has attracted tremendous attention.

Gas chromatography-mass spectrometry (GC-MS) is the earliest chromatography-mass spectrometry technology which is mostly well-developed. The chromatographic part of the GC-MS includes a sample injector (either manual or automatic sampler), vaporization chamber, column oven, and carrier gas system. Depending on the amount of sample, a split or splitless injection method can be used. Then the multi-component sample is injected into the chromatographic column. Due to the different interactions between the different components and the stationary phase of the chromatographic column, after a certain period, the components are separated from each other and enter into the mass spectrometer successively.

It is worth noting that the column outlet is at atmospheric pressure while the mass spectrometer operates at a high vacuum. Therefore, if a packed column is used, an interface (such as a molecular separator) is required to remove the carrier gas in the column effluent as much as possible. This experiment uses a capillary column that can be inserted directly into the ion source of the mass spectrometer.

The mass spectrometry part of the GC-MS includes an ion source, a mass analyzer, and a detector. In this experiment, an electron impact (EI) ion source is used. Under 70eV electron bombardment, the neutral sample molecules lose electrons with low ionization energy to become the charged molecular ions, and then further breakage of chemical bonds occurs, resulting in low-mass fragment ions. The function of the mass analyzer is to separate the ions generated by the ion source according to the value of m/z and finally be measured by the detector. There are many types of mass analyzers, and a quadrupole mass analyzer is used in this experiment. Such an analyzer consists of four parallel cylindrical metal electrodes, and the opposite electrodes are connected diagonally to form two sets of electrodes. When a direct current (DC) voltage U_{dc} and a radio frequency alternating current (AC) voltage V_{rf} with equal value and opposite direction are applied between the two sets of electrodes, a hyperbolic electric field would be generated in the space enclosed by the quadrupole. When U_{dc}/V_{rf} is constant, only m/z ions with a certain value can stably oscillate through the quadrupole rod and then reach the detector. The other ions would hit the quadrupole and then be pumped away by the vacuum system. The DC voltage U_{dc} and the radio frequency AC voltage V_{rf} can be changed to realize the mass scanning for obtaining a mass spectrum. Ascribed to the small size and fast scanning speed, the quadrupole mass analyzer is suitable for chromatography-mass spectrometry.

Equipment and reagents

(1) **Equipment.** Agilent 7890B-5977B GC-MS, Restek Rtx-5 MS capillary column, small ultrasonicator.

(2) **Reagents.** Chromatographically pure n-hexane, methanol, sodium hydroxide, anhydrous sodium sulfate, pH test paper, rapeseed oil.

Procedures

(1) **Sample processing.** Weigh 2mg of rapeseed oil into a clean 5mL glass bottle, add 100μL of methanol (containing 0.5mol/L NaOH) and 1mL of n-hexane, then cover the bottle. After 1 min of sonication, take out the bottle and placed it in the ice bath. After the temperature drops and the two phases are completely separated, use a glass pipette to take the upper layer of n-hexane and wash it with ultrapure water, until it becomes neutral. Add anhydrous sodium sulfate to remove trace water in the n-hexane sample layer, and finally transfer the treated n-hexane to the injection bottle for further analysis.

(2) **Chromatographic conditions setting.** Due to the high fatty acid content in rapeseed oil, the injection method is set as a split injection with a split ratio of 100 (this ratio depends on the different samples). According to the nature of the sample, set the parameters such as inlet temperature, carrier gas flow rate, programmed temperature, and so on.

(3) **Setting conditions of mass spectrometry.** Set the ion source temperature, interface temperature, detector voltage, scanning mode (full scan or selected ion scan), mass range, etc. Pay attention to setting the solvent removal time.

(4) **Sample determination.** Select manual or automatic injection and the injection volume is generally 1μL. After sampling, set the instrument on the standby state.

Data processing

(1) Spectral analysis. Analyze the generation mechanism of the main fragment ion peaks and the characteristics of isotopic ion peak clusters in the mass spectrum to derive the elemental composition and molecular structure of typical fatty acid methyl esters.

(2) NIST database searching. Search the obtained mass spectrum against the NIST database, and compare the difference and matching degree between the experimentally obtained and the standard mass spectrum.

Notes

(1) After the rapeseed oil is methylated and derivatized under alkaline conditions, the solution becomes alkaline. It needs to be washed to neutrality and the residual water should be removed with anhydrous sodium sulfate. Otherwise, it would cause the loss of the stationary phase in the chromatographic column.

(2) As the fatty acid content in rapeseed oil is high, if the split injection method is not chosen, the current will be too large, which may burn out the filament and even damage the detector.

Questions

(1) Please explain the reasons for setting the solvent removal time.

(2) What is the significant difference between the mass spectra of saturated and unsaturated fatty acid methyl ester? Can the position of the double bond be identified by mass spectrometry?

(3) Can *cis*-and *trans*-fatty acid methyl esters be distinguished by chromatography-mass spectrometry?

(4) Please deduce the generation mechanism of the fragment ion peak with m/z value of 74 in the mass spectrum of saturated fatty acid methyl ester.

Exp. 39 Qualitative and Quantitative Analysis of Baicalin in Radix Scutellariae by Liquid Chromatography-Mass Spectrometry

Objectives

(1) To master the analytical methods of liquid chromatography-mass spectrometry (LC-MS).

(2) To be familiar with the main components of LC-MS.

Principles

Liquid chromatography-mass spectrometry (LC-MS) is another widely used chromatography-mass spectrometry technique. LC-MS are mainly composed of a liquid chromatography system, a connection interface, a mass analyzer and a computer data processing system. The relative mature interface and ionization technologies of LC-MS are electrospray ionization (ESI) and atmospheric pressure chemical ionization (APCI).

Baicalin, one of the main active ingredients of Radix Scutellariae, is a flavonoid compound with the molecular formula of $C_{21}H_{18}O_{11}$. Its molecular structure is demonstrated as follows (Figure 3-9):

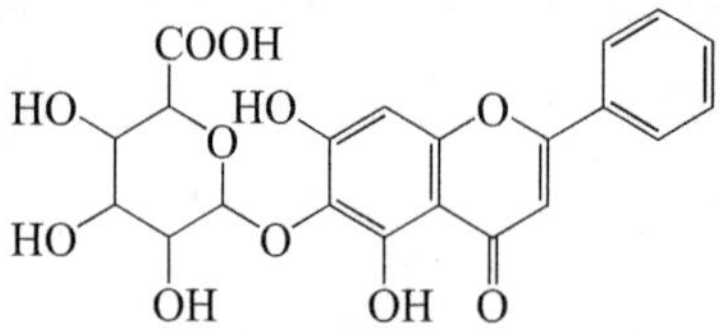

Figure 3-9   The molecular structure of baicalin

In this experiment, LC-MS is applied to identify the chemical structure and determine the content of baicalin in Radix Scutellariae. It has the characteristics of sensitivity, accuracy, and speed.

Equipment and reagents

(1) **Equipment.** Agilent 6230 TOF LC/MS hybrid system, Agilent ChemStation, analytical balance, sonicator, $0.45\mu m$ membrane filters, microsyringe.

(2) **Reagents.** Radix Scutellariae, baicalin (standard, 99%), methanol (GR), aceto-

nitrile (GR), formic acid (AR), ultrapure water.

Procedures

(1) Operating conditions (reference values)

① Chromatographic conditions. Column: Waters Symmetry Shield RP-18 (100mm × 2.1mm × 3.5μm) with protecting column C_{18} (10mm × 2.1 mm × 3.5μm). The mobile phase is solvent A (water 0.1% formic acid) + solvent B (acetonitrile, 0.1% formic acid), the flow rate is 0.4mL/min and the injection volume is 5μL. Elution gradient (A+B=100%) is listed in Table 3-14.

Table 3-14 Elution gradient for baicalin determination

t/min	0.00	2.00	6.00	11.00	11.25	12.25	12.50
[B]/ %	30	30	60	85	99	99	30

② MS conditions. Ion source is an electrospray ionization source (ESI). Ion source ejection voltage is 4kV, the flow rate of drying gas (N_2) is 11.5L/min. The drying gas temperature is 350℃, nebulizing gas (N_2) pressure is 2.4×10^5 Pa, the m/z of precursor ion is 447.2 and the production is 271.1.

(2) Sample preparation

① Standard stock solution. Accurately weigh about 10mg of baicalin reference substance, dissolve it with methanol and transfer it to a 100mL volumetric flask, dilute to the mark, and mix well.

② Standard solutions. Accurately transfer appropriate volumes of the stock solution into the volumetric flasks and dilute them with methanol-water (1 : 1, volume ratio) to the mark to prepare a series of standard solutions with the concentrations at about 0.5μg/mL, 2.5μg/mL, 10μg/mL, 20μg/mL.

③ Sample solution. Grind the Radix Scutellariae (50g) into powder, weigh about 0.1g into the beaker, add a certain amount of methanol-water mixture (1 : 1, volume ratio) into it and sonicate the beaker for 30 min to dissolve the powder. Transfer the solution into a 100mL volumetric flask. After the solution cools down to room temperature, dilute it with a methanol-water mixture (1 : 1, volume ratio) to the mark and mix well. Accurately transfer 5.00 mL of the above solution into a 50 mL volumetric flask, add methanol-water (1 : 1, volume ratio) to dilute it to the scale line, mix well and filter it with a 0.45μm membrane filter.

(3) Samples analysis.
Separately inject 5μL of the standard solutions and sample solution into the LC-MS equipment, and record the total ions chromatogram (TIC), mass spectrum and chromatogram of baicalin. Repeat the measurement 3 times.

Data processing

(1) Qualitative analysis. Under the positive ion detection mode, the quasi-molecular ion peak of baicalin (relative molecular weight 446.35) can be detected and the m/z of [M+

$H]^+$ is 447.2. The detected characteristic ion with m/z of 271.1 is the daughter ion of $[M+H]^+$, due to the loss of glucuronic acid. The identification of baicalin can be finally determined according to the mass spectrum and the chromatographic performances.

(2) Quantitative analysis

① Construct a calibration curve by plotting the peak area against the concentration of baicalin and obtain the regression equation and correlation coefficient. Fill in the data in Table 3-15.

② Calculate the content of baicalin in Radix Scutellariae with the peak area of baicalin in the regression equation (Table 3-16).

Table 3-15   Retention time and peak area of different concentrations of baicalin

Concentration of baicalin/(mg/mL)	Retention time, t_R/min	Peak area
Regression equation		
Correlation coefficient		

Table 3-16   Baicalin content in Radix Scutellariae

Sample	Retention time, t_R/min	Peak area	Content of baicalin/%	Average content/%
1				
2				
3				

Notes

Compared with GC-MS, the system noise of LC-MS is large, and measures such as purification of organic solvents, purification of samples, and regular cleaning of the system should be taken to reduce the impact of noise.

Questions

(1) What are the components of LC-MS? How many types of interfaces are used in LC-MS?

(2) What are the characteristics of LC-MS compared with GC-MS?

Exp. 40　Paper Chromatography Identification of Amino Acids

Objectives

(1) To grasp the principle of paper chromatography on separation and identification of mixtures.

(2) To grasp the proper operation method of paper chromatography.

Principles

Chromatography can be divided into column chromatography, paper chromatography and thin layer chromatography (TLC) according to different types of stationary phases. Paper chromatography uses filter paper as the carrier, water adsorbed on the paper as the stationary phase, and the organic solvent that is incompatible with water as the mobile phase. Paper chromatography is mainly used in the analysis of polar compounds such as sugar and amino acids. During chromatography, the sample is spotted at about 2 to 3cm from the end of the filter paper, which is called the origin. In a closed container, the chromatography solvent then penetrates the paper by capillary action and, passing over the sample spot, carries along with the various components of the sample. The components move with the flowing solvent at velocities that are dependent on their solubilities in the stationary and flowing solvents. Due to different partition coefficients (K), components appear at different positions on the filter paper. The position of individual species on the paper chromatogram can be expressed as relative front or retardation factor (R_f), which is defined as:

$$R_f = \frac{\text{distance between the origin and the center of the spot}}{\text{distance between the origin and the solvent front}}$$

Under the same experimental conditions, the R_f value stays constant. Therefore, it can be used for qualitative analysis of substances. In this experiment, the mixture of n-butanol-acetic acid-water (4 : 1 : 1) (BAW system) is used as the developing solvent, and two amino acids glycine (NH_2CH_2COOH) and methionine ($CH_3SCH_2CH_2CH(NH_2)COOH$) will be separated by ascending paper chromatography. Due to the difference in the structure of the two, glycine is more polar than methionine. Thus, the mobility of glycine is slower on the filter paper, resulting in a smaller R_f value. After developing, the amino acids undergo a chromogenic reaction with ninhydrin at 60℃, forming purplish-red spots on the chromatography paper.

Equipment and reagents

(1) Equipment. Chromatography tank with a lid, chromatography paper (or filter paper), capillary tubes, a drying oven.

(2) Reagents. 0.5% glycine standard solution, 0.5% methionine standard solution, the mixture of an equal amount of 0.5% glycine and 0.5% methionine. The developing solvent is n-butanol-acetic acid-water (4 : 1 : 1) and the chromogenic spray reagent is 1‰ ninhydrin in alcohol.

Procedures

(1) **Spotting.** Take a piece of chromatography paper (15cm×5cm) and punch a small hole in the middle at 1cm from the top. Gently draw a straight line with a pencil 1.5 to 2 cm from the lower end of the chromatographic paper. Spot the above standard solution and sample mixture solution 3 to 4 times with a capillary tube to make the diameter of the spot about 2mm, and dry them in air.

(2) **Developing.** Add 35mL of developing solvent into the dry chromatographic tank. Hang the spotted filter paper vertically in the chromatography tank, and close the tank lid to saturate the paper for 10 minutes. Then dip the paper's edge into the solvent about 0.3-0.5cm and develop it. When the solvent front is developed to a suitable position (about 12 cm), take out the filter paper with forceps and immediately mark the position of the solvent front line.

(3) **Coloration.** Dry the filter paper, spray it with ninhydrin solution, place it in a 60℃ oven for 5 minutes, or carefully heat it above the electric furnace to visualize purple-red spots.

Data Processing

Mark the range of each spot, find out the spot's center, and calculate the R_f values of each sample. Identify the mixed sample.

Notes

(1) Prepare the developing solvent in advance and shake it up completely.

(2) Use the new capillary tubes to avoid cross-contamination.

(3) Ninhydrin solution can also color body fluids such as sweat, and forceps should be used to remove the filter paper to avoid filter paper contamination.

(4) Ninhydrin solution should be prepared before use and refrigerated for further use.

(5) Spray the chromogenic reagent evenly and appropriately.

Questions

(1) What factors affect the R_f value?

(2) How to get a neat paper chromatogram with concentrated spots and a consistent solvent front?

(3) Why must the filter paper be saturated in a chromatography tank before it is dipped into the developing solvent? What is the requirement for time and temperature of saturation?

(4) What the advantages and limitations of paper chromatography?

Exp. 41   The Identification of Rhizoma Coptidis by Thin Layer Chromatography

Objectives

(1) To master the preparation method of TLC plate.

(2) To master the basic operation of TLC.

(3) To understand the application of TLC in the identification of traditional Chinese medicine.

Principles

Thin layer chromatography (TLC) is a chromatographic technique that applies the stationary phase to a glass or plastic plate and separates the components of a mixture due to different affinities between the stationary phase and mobile phase after the sample is spotted and developed. It is widely used in separating multicomponent pharmaceutical formulations.

Rhizoma Coptidis is traditional Chinese medicine and is widely used for its antibiosis efficacy. It contains a variety of alkaloids, of which berberine hydrochloride is the main active ingredient, which can emit yellow fluorescence when exposed to ultraviolet light (365nm). Alkaloids contained in Rhizoma Coptidis can be successfully separated by TLC method, and the Rhizoma Coptidis test drug can then be identified by comparing it with the reference drug and berberine hydrochloride reference substance.

Equipment and reagents

(1) **Equipment.** UV analyzer, oven, glass plate (5cm × 10cm), mortar, double-groove chromatographic chamber, medicine spoon, capillary tube, pencil, ruler.

(2) **Reagents.** Rhizoma Coptidis test drug, Rhizoma Coptidis reference drug, berberine hydrochloride reference substance, benzene (AR), acetoacetate (AR), isopropanol (AR), methanol (AR), concentrated ammonia (CP), sodium carboxymethylcellulose in water (CMC-Na, 0.7%), silica-gel G (for TLC).

Procedures

(1) **Preparation of TLC plates.** Weigh 10g of silica gel G into a mortar, add approximately 25mL of CMC-Na solution, and then grind it together in one direction until no air bubbles are present, resulting in a uniform suspension. Transfer an adequate amount of the suspension onto a cleaned and dried glass plate with a medicine spoon, spread it uniformly, and let it dry in air. Before being used, activate it at 105°C for 30min and then store it in a desiccator for cooling.

(2) **Preparation of the samples.** Weigh 50mg powder of Rhizoma Coptidis test drug, add 5mL of hydrochloric acid-methanol (1 : 100). Heat and reflux it for 15min. After filtration,

add the solvent to 5mL as the test drug solution. Prepare a solution of Rhizoma Coptidis reference drug in the same manner as the reference drug solution. Dissolve the reference substance of berberine hydrochloride in methanol to prepare a solution with the concentration of 0. 5mg/mL as the reference solution.

(3) **Spotting.** Using a pencil, gently draw a starting line about 1. 5cm away from the bottom edge of the plate and a solvent front line 1. 0cm away from the top edge. Apply 1μL of each of the above three solutions separately to the starting line by capillary tubes.

(4) **Developing.** Take a mixture of benzene-acetoacetate-isopropanol-methanol-water (6 : 3 : 1. 5 : 1. 5 : 0. 3) as the developing solvent. Put the plate in a chamber pre-equilibrated with the vapor of concentrated ammonia, and then develop it. When the solvent front reaches the solvent front line draw in advance, and remove the plate.

(5) **Visual inspection.** Allow the plate to dry for a few minutes, and then observe it under ultraviolet light (365nm). Circle the principal spots that are illuminated with a pencil.

Data processing

Compare the color and position of the fluorescence spots obtained from the three chromatographic bands. The fluorescence and the position of the principal spots in the chromatogram obtained with the reference solution and test solution of Rhizoma Coptidis drugs should be identical, and also there should be a same yellow spot in the chromatogram of the test drug solution and the solution of berberine hydrochloride reference.

Notes

(1) The surface of the plates should be uniform, flat, and smooth without pockmarks and bubbles. The thickness of the layer is generally 0. 2-0. 3mm. Before being activated, the plate should be dried naturally to prevent cracking.

(2) After being activated, the thin layer plate should be placed in a desiccator to cool immediately after activation in the oven.

(3) The amount of spots should be appropriate. Too little sample will produce indistinct spots, while excessive sample will cause tailing. When applying the samples, be careful to avoid puncturing the surface of the TLC plate. The diameter of the circular spots is generally less than 2-3 mm.

(4) Spots intervals should be adjusted according to the diffusion of the spots and with no interference between neighboring spots, usually, not less than 8mm.

(5) Seal up the chamber to avoid evaporation of the solvent. Otherwise, the ratio of the developing agents will be changed and the separation quality will be affected.

Questions

(1) How many methods are mainly used to color the solute in TLC?

(2) Under what conditions can a substance emit fluorescence?

(3) Why should the chromatographic chamber be saturated by ammonia vapor while developing?

Exp. 42 Determination of Anions in Tap Water by Ion Chromatography

Objectives

(1) To understand the basic principles and operation skills of ion chromatography.

(2) To master the qualitative and quantitative analysis methods of ion chromatography.

Principles

Ion chromatography (IC) is a liquid chromatography method derived from ion exchange chromatography that combines the efficient separation of chromatography with the automatic detection of ions. IC uses ion exchange resin as the stationary phase and electrolyte solution as the mobile phase, and usually uses a conductivity detector for detection. IC is available in single-column and double-column versions, which generally consist of four parts, namely a conveying system, a separation system, a detection system, and a data processing system.

In this experiment, the anion exchange resin was used as the stationary phase and the $NaHCO_3$-Na_2CO_3 mixture as the eluent to analyze the three anions of Br^-, NO_3^- and SO_4^{2-} in the water. When the test solution containing the anions to be measured enters the separation column, the following exchange process occurs in the separation column:

$$RHCO_3 + MX \rightleftharpoons RX + MHCO_3$$

Wherein R represents an ion exchange resin.

Because the eluate continuously flows through the separation column, the various anions exchanged on the anion exchange resin are eluted again, and the elution process occurs. Due to their different affinity with the ion exchange resin, the exchange and elution process of various anions is different. The ions with small affinity first elute out of the separation column, and the ions with large affinity later flow out of the column. That is how different ions are separated.

When the anions to be measured are eluted from the column and enter the conductivity cell, the conductivity detector is required to detect the change in the conductivity in the eluate at any time. However, because the concentration of HCO_3^- and CO_3^{2-} in the eluate is much larger than the concentration of anions in the sample, the conductivity contribution of the test solution ions is insignificant compared with the conductivity value of the eluent itself. Therefore, the conductivity detector is difficult to detect the conductivity change caused by the change in ion concentration of the test solution. For ion chromatography with an inhibition column, H^+ from the regeneration solution enters the eluent through the cation exchange membrane and binds to CO_3^{2-}, HCO_3^- and X^- of the eluent to form weakly ionized H_2CO_3 and strongly ionized HX. To maintain the electroneutrality of the eluent and the regeneration solution, the stoichiometric Na^+ moves in the opposite direction, that is, from the eluent channel to the regeneration solution, and finally is brought into the waste liquid, resulting in the conversion of $NaHCO_3$ and Na_2CO_3 in the eluate to H_2CO_3, which greatly reduces the background conductance. While the MX in the specimen is converted to the corresponding acid HX, the ion mobility of H^+ is 7 times that of metal ion M^+, so the determination of ion conductance in the test solution can be realized.

Equipment and reagents

(1) Equipment. Metrohm type 861 ion chromatograph, IC Net 2.3 chromatography workstation, Metrosep A supp 4 anionic exchange column (250mm × 4.0mm i.d.), Metrohm MSM Ⅱ suppressor+853 Type CO_2 suppressor, conductivity detector.

(2) Reagents

① Weigh 19.10g Na_2CO_3 (AR) and 14.30g $NaHCO_3$ (AR) (both have been baked in a 105℃ oven for 2h and cooled to room temperature), and dissolve them in high-purity water. Transfer into a 1000 mL volumetric flask, add water to the mark and shake well. This rinse stock solution is then stored in a polyethylene bottle and refrigerated. This eluent stock solution is 0.18mol/L Na_2CO_3 + 0.17mol/L $NaHCO_3$. ② Prepare 1000mg/L Br^-, 1000 mg/L NO_3^-, 1000mg/L SO_4^{2-}, dilute them as the standard solution during determination. The standard mixture solution contains 20mg/L Br^-, 20mg/L NO_3^- and 200mg/L SO_4^{2-}.

Procedures

(1) Preparation of anionic eluent. Pipette 0.18mol/L Na_2CO_3 + 0.17mol/L $NaHCO_3$ anion elucidation stock solution 10.00mL, diluted with high purity water to 1000mL, shake well. This eluent is 1.8mmol/L Na_2CO_3 + 1.7mmol/L $NaHCO_3$.

(2) Machine startup. Turn on the power switch of ion chromatograph and IC Net 2.3 chromatography workstation. Start the pump, adjust the mobile phase flow rate to 1mL/min and make the system balance for 30min. Wait for the instrument to stabilize until the baseline of the chromatographic elute curve is straight.

(3) Single standard solution analysis. Adjust the instrument to the injection mode, start the Fill key, and inject 1mL of each anion standard solution using the syringe. Then start the Inject key, start the chromatographic analysis, and record the retention time of each anion after the peaks are all out.

(4) Standard mixture solution analysis. Following step (3), directly inject the mixed anion standard onto the pump, and identify the retention time of each anion in the mixture.

(5) Plotting the working curve. Pipette 0.50mL, 1.00mL, 2.00mL, 3.00mL, 4.00mL of the anion mixture standard in five 10mL volumetric flasks, dilute with high-purity water to the mark and shake well. Inject each solution onto the column twice, and record the chromatogram. Plot the peak area against the ion concentration and obtain the working curve of each ion.

(6) Tap water analysis. Take a laboratory tap water sample, filter it through a 0.45μm membrane, and repeat the injection twice under the same experimental conditions to record the chromatogram. Identify the ions in tap water by the retention time and quantify them by the peak area.

Questions

(1) Briefly describe the separation mechanism of ion chromatography.

(2) Why is it necessary to add a suppressor in front of the conductivity detector?

Exp. 43　Chiral Separation of Ofloxacin Enantiomer by Capillary Zone Electrophoresis

Objectives

(1) To be familiar with the fundamental principle and method of capillary zone electrophoresis (CZE).

(2) To know the application of CZE in the separation of chiral drugs.

Principles

Capillary electrophoresis (CE) is a new type of liquid phase separation technology. It applies a high-voltage direct current field as the driving force and uses capillary as the separation channel to achieve sample separation according to the difference in the mobility and distribution behavior of each component in the sample. Due to its advantages of high efficiency, rapidity and micro-quantity, it has attracted great attention in the field of chiral analysis in recent years.

Ofloxacin (ofloxacin, OFLX), a kind of common quinolone antibiotic, is available in the form of the racemate and levofloxacin for clinical use. The antibacterial activity of S-($-$)-ofloxacin is $8 \sim 128$ times higher than that of R-($+$)-ofloxacin and twice of the racemate. The antibacterial mechanism is to hinder the replication of bacterial DNA by inhibiting the activity of bacterial DNA rotational enzyme (bacterial topoisomerase II.). The chemical structural formula of levofloxacin is as follows (Figure 3-10):

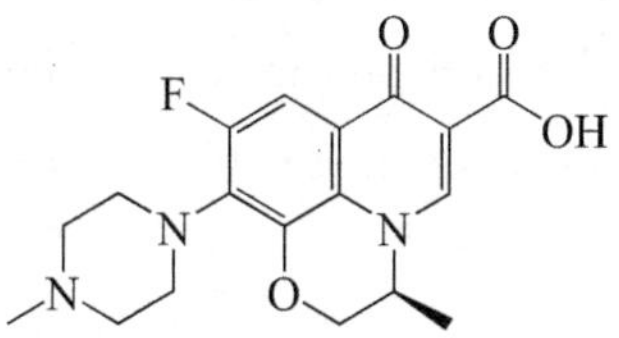

Figure 3-10　Molecular structure of levofloxacin

In this experiment, dimethyl-β-cyclodextrin (DM-β-CD) is used as a chiral selector to separate the enantiomers of ofloxacin by CZE.

Equipment and reagents

(1) **Equipment.** Capillary electrophoresis, pH meter, fused silica capillary, analytic balance, $0.45\mu m$ membrane filter.

(2) **Electrophoresis conditions (reference value).** The concentration of cyclodextrin is 40mmol/L. The electrophoresis buffer is 70mmol/L KH_2PO_4 (H_3PO_4 adjust pH to 2.5). The applied voltage is 20kV, the detection wavelength is 280nm, and the capillary temperature is 25℃.

(3) **Reagents.** Ofloxacin, levofloxacin, dimethyl-β-cyclodextrin (DM-β-CD, AR), KH_2PO_4 (AR), H_3PO_4 (AR), deionized water, 0.1mol/L NaOH solution.

Procedures

(1) **Preparation of supporting electrolyte solution.** Prepare 70mmol/L KH_2PO_4 solution containing 40mmol/L dimethyl-β-cyclodextrin and adjust the pH to 2.5 with H_3PO_4 solution. Filter the solution with a $0.45\mu m$ membrane and ultrasonically degas it for further use.

(2) **Preparation of sample solutions.** Accurately weigh the appropriate amount of ofloxacin and levofloxacin, respectively. Dissolve them in deionized water and make their final concentration to be 1.4mg/mL. Filter the solution with a $0.45\mu m$ membrane and ultrasonically degas it for further use.

(3) **Injection and separation.** Before analysis, rinse the capillary with 0.1mol/L NaOH solution and deionized water for 10min, respectively. Then rinse it with background electrolyte buffer solution for 5min and finally balance it under the applied voltage for 10min. Inject ofloxacin (1.4kPa, 1s) to get the electrophoretogram. Add levofloxacin to further confirm the peak position of levofloxacin and dextrofloxacin. Inject the levofloxacin sample and get the electrophoretogram (13.8kPa, 1s).

Data processing

(1) Record chiral separation electrophoretograms of ofloxacin and levofloxacin.

(2) Based on the area normalization method, calculate the content of *d*-and *l*-isomers in ofloxacin and levofloxacin samples, respectively.

Notes

(1) In CE experiment, some important factors such as type, the concentration of cyclodextrin, concentration of electrolyte, column temperature and voltage which can affect separating should be optimized for improving resolving power.

(2) There will be the best-separating effect when the concentration of DM-β-CD is between 30-40mmol/L. Therefore, it should be controlled in the range of 30-40mmol/L.

Questions

(1) What are the separating mechanisms in CE?

(2) How to ensure the good reproducibility of migration time in CE?

(3) What factors can affect the bandwidth in CE?

参考文献

[1] 华中师范大学，东北师范大学，陕西师范大学，等.分析化学实验 [M].4 版.北京：高等教育出版社，2015.

[2] 王新宏.分析化学实验（双语版）[M].北京：科学出版社，2009.

[3] 严拯宇，杜迎翔.分析化学实验与指导 [M].3 版.北京：中国医药科技出版社，2015.

[4] 吴性良，朱万森.仪器分析实验 [M].2 版.上海：复旦大学出版社，2008.

[5] 袁霖，李中燕，袁先友，等.仪器分析实验 [M].长沙：中南大学出版社，2019.

[6] 张晓凤，柏俊杰，曹坤，等.现代仪器分析实验 [M].重庆：重庆大学出版社，2020.

[7] 孙尔康，张剑荣.仪器分析实验 [M].2 版.南京：南京大学出版社，2015.

[8] 柳仁民.仪器分析实验 [M].2 版.青岛：中国海洋大学出版社，2013.

[9] 武汉大学.分析化学（下册）[M].6 版.北京：高等教育出版社，2018.

[10] 华中师范大学，东北师范大学，陕西师范大学，等.分析化学（下册）[M].4 版.北京：高等教育出版社，2011.

[11] Harris D C. Quantitative Chemical Analysis [M]. 8th ed. New York：W. H. Freeman and Company，2009.

[12] Hage D S，Carr J D.分析化学和定量分析（英文版）[M].北京：机械工业出版社，2012.

[13] Ieggli C V S，Bohrer D，Noremberg S，et al. Surfactant/oil/water system for the determination of selenium in eggs by graphite furnace atomic absorption spectrometry [J]. Spectrochimica Acta Part B，2009，64：605-609.

[14] 李鸿，蒋越华，秦玉燕，等.原子荧光光谱（AFS）和石墨炉原子吸收光谱（GFAAS）法测定富硒粮食中硒含量 [J].中国无机分析化学，2021，11（3）：89-93.